Practical Astronomy

Springer
London
Berlin
Heidelberg
New York
Barcelona
Hong Kong
Milan
Paris
Santa Clara
Singapore
Tokyo

Other titles in this series

The Observational Amateur Astronomer
Patrick Moore (Ed.)

The Modern Amateur Astronomer
Patrick Moore (Ed.)

Telescopes and Techniques
Chris Kitchin

Small Astronomical Observatories
Patrick Moore (Ed.)

The Art and Science of CCD Astronomy
David Ratledge (Ed.)

The Observer's Year
Patrick Moore

Seeing Stars
Chris Kitchin and Robert W. Forrest

Photo-guide to the Constellations
Chris Kitchin

The Sun in Eclipse
Michael Maunder and Patrick Moore

Software and Data for Practical Astronomers
David Ratledge

Observing Meteors, Comets, Supernovae
and other Transient Phenomena
Neil Bone

Astronomical Equipment for Amateurs
Martin Mobberley

Stephen F. Tonkin (Ed.)

Stephen F. Tonkin, BSc, PGCE

Cover photograph: CCD image by Andy Saulietis, Danciger Telescope Group

ISBN 1-85233-000-7 Springer-Verlag London Berlin Heidelberg

British Library Cataloguing in Publication Data
Amateur telescope making. - (Practical astronomy)
1.Telescopes - Design and construction - Amateurs' manuals
I.Tonkin, Stephen F.
681.4′123
ISBN 1852330007

Library of Congress Cataloging-in-Publication Data
Amateur telescope making / Stephen F. Tonkin (ed.).
p. cm. - (Practical astronomy)
Includes bibliographical references and index.
ISBN 1-85233-000-7 (pbk. : alk. paper)
1. Telescopes-Design and construction-Amateur's manuals.
I. Tonkin, Stephen F., 1950- . II. Series.
QB88.A622 1998 98-33566
681′.4123-dc21 CIP

Printed in Great Britain
3rd printing 1999

Typeset by EXPO Holdings, Malaysia
Printed at the University Press, Cambridge
58/3830-5432 Printed on acid-free paper SPIN 10754588

We dedicate this book, with humility, to the memory of the late Tom Waineo. Tom epitomised the most laudable qualities of the amateur telescope maker, giving selflessly of his time and using his extensive experience and wisdom in order to help others. We hope that our efforts will help to keep Tom's star shining.

Contents

IV Astrophotography

V Appendices

Introduction

Over the last few decades the range of mass-produced equipment available to amateur astronomers has increased in both extent and capability, and decreased in real-term cost. Obvious examples of the enhanced capability of amateur equipment lie in CCD cameras and computer-controlled mounts. The CCD is said to increase the light-gathering power of a telescope by a factor of about a hundred; that is, it is possible to take images with an 8 in (20 cm) instrument that would previously have required an 80 in (2 m) telescope using photographic emulsion. Of course, the resolving power of the 8 in is not also increased! Computer control enables simplified finding of faint objects and, coupled with a CCD, can automatically guide the telescope during imaging. Where then, in this context, is the place for amateur telescope making (ATMing) and the basement tinkerer, the person Albert Ingalls referred to, in his *Amateur Telescope Making* trilogy, as the TN – the Telescope Nut?

Quite simply, the opportunities for TNs – who are now better known as ATMs (amateur telescope makers) – have also increased correspondingly, and their craft has developed far beyond what can legitimately be termed "basement tinkering". Richard Berry's *CCD Camera Cookbook* has resulted in the construction and use of hundreds of home-made CCD cameras, at a cost well below that of the commercially available instruments. As Al Kelly demonstrates (Chapter 14), making a *Cookbook* CCD is, while time-consuming, not a particularly difficult task, and the resulting images easily rival those taken with the mass-produced products. Similarly, the innovations of Mel Bartels and others (Chapter 10) have extended computer control to the most common ATM telescope, the Dobsonian-mounted Newtonian, again at a much-reduced cost compared with commercial offerings.

While cost reduction has always been one of the driving forces behind ATM, it is neither the only one nor the most powerful. However, there are still among us those who aim to make good astronomical kit at a shoestring cost, and there are several such examples in the first section of this book. You should understand that the costs are cut not by sacrificing optical quality, but usually by adapting free or low-cost items that were intended for another purpose. While we don't eschew good craftsmanship, we stand by the principle (or is it merely an excuse?) that telescopes are primarily for looking *through*, not looking *at*. As long as the telescope holds the optical components rigidly in collimation while excluding stray light from the eyepiece, and the mount is as steady as the Rock of Gibraltar, while permitting smooth movement about two axes of rotation, all else is optional.

More often, the ATMing impulse is a response to mass production. One of the consequences of this mass production is standardisation, as witnessed by the ubiquitous 8 in (20 cm) Schmidt–Cassegrain telescopes. If you want something non-standard, you are often left with two choices: either to forgo the cost advantage of mass production by having a one-off instrument specially made, or to make the instrument yourself. The section "Specialised Telescopes" shows how some ATMs have met this need. This is the realm where much innovation takes place, and some ATMs have found their developments to be so popular that they have gone into commercial production.

Another powerful stimulus to ATMing lies in human nature – some of us are inveterate tinkerers. If we buy a telescope, within days we find we have invalidated the warranty. A few days (or, in some severe cases, hours) later, the first modification is made and within a month or so the instrument bears little relation to the original product. Our workshops usually contain several telescopes, mostly in various states of dismantlement, and we are known in astronomical circles as people who spend more time tinkering with telescopes than actually using them. This ailment is probably incurable, but its cravings are certainly satisfied by ATMing!

Whatever the impulse that attracted you to this book, you will find a number of ways that it differs from most other books on the subject. The most obvious of these is that each chapter is written by someone who has, to some extent, become an expert in the realm his chapter covers and who has, in most cases, spent considerable time helping others to attain a similar level of expertise. Each contributor is someone with a proven ability to make equipment that works effectively, and many have devised creative ways of using common artefacts – this inventiveness will inspire you to do likewise. He is also someone who is willing to help you further, should the need arise. To this end, the publishers have dedicated a World Wide Web site to this book, via which you can contact any of the contributors to the book.

This linking to a web site also enables the book to be kept up to date. In particular, software is continuously under development, but the latest versions may be obtained on the web.

The multi-contributor nature of the book is a microcosm of another aspect of ATMing: that of mutual aid. Most ATMs are eager to share the experiential fruits of their work with others, and many of the international group of contributors to this book "met" on the Internet via the ATM Mailing List, which was established purely to facilitate this sharing. Many of us have had the privilege of being advised, via this medium, by experts in their field. Those of us who began our ATMing with no tutor but a book will appreciate the value of a resource that can be questioned when the need arises; the ATM Mailing List is just such a resource and is an excellent forum for sounding out any ideas that this book may inspire.

Another difference is that optical work is not specifically covered. There are several reasons for this. It requires an entire book to itself, and there are already several excellent publications on the subject available. While it is true that optical work is a craft that can be learned, there are relatively few excellent amateur opticians – like any other craft, it takes time and practice to achieve excellence. Consequently many, but by no means all, first amateur mirrors are of inferior quality and, unless you have a specific desire to develop the necessary skills, it often makes good sense to purchase optical components from a reputable source. There usually is little financial saving, if any, to be made by making a small primary mirror, although the skills so gained will prepare you for those specialised tasks where there are significant savings are to be made. Obviously, most specialised instruments require specialised optical components, and for these you may have no choice but to make them yourself. If this is the case, it is advisable to learn your skills on a "standard" mirror, such as a 6 in (15 cm) f/8 or an 8 in (20 cm) f/7, before attempting the specialised optical surfaces.

Whether you decide to buy your optics or to make them yourself, it is essential to learn to test them. It is unfortunate but true that not all mass-produced optical components are of the quality that the maker claims for them, and it makes excellent sense to be able to evaluate them for yourself. In any case, you will need to test your finished telescope, whatever the source of the optical components.

The Bibliography lists several excellent books on optical design, work and testing.

In any book of this nature that has an international body of contributors, it is inevitable that there will be a "confusion" of units of measurement. Even in countries that still use non-metric measures, focal lengths, particularly of eyepieces, are usually expressed in millimetres. The style I have adopted is that the author's chosen system of units is given first, followed by an appropriate "translation", where this is possible, when the measurement is first used. I have attempted to render these as translations into standard sizes where this is appropriate. For example, $\frac{3}{4}$ in is almost exactly 19 mm, but the nearest metric equivalent to $\frac{3}{4}$ in thick plywood is 18 mm thick. There are some instances where a sensible translation is not helpful, usually in relation to screw threads. For example, a tripod bush for a camera has a $\frac{1}{4}$ in UNC ($\frac{1}{4}$ in 20 tpi) thread – there is no metric equivalent. In these instances, I have not given equivalents.

Finally, although detailed instructions are provided for many of the projects in this book, a hallmark of ATMing is creativity. Each project will perform well if it is made as the author made it, but most are capable of adaptation and development to your specific needs or to the materials you have available. Although the projects vary greatly in simplicity/difficulty of construction, most will fall within the capability of an ATM with a moderately well-equipped workshop and reasonable workshop skills. Whether you use this book as a manual or as a source of ideas that you will develop to meet your own particular requirements, I hope you will find it as stimulating to read and use as I have found it to compile.

Stephen Tonkin
April, 1998

Part I

Shoestring Telescopes

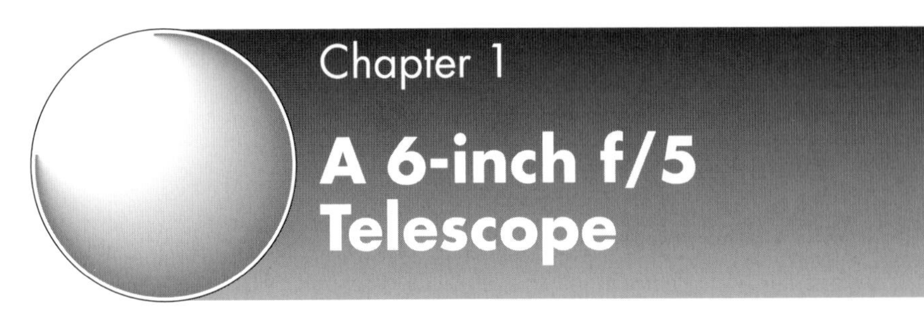

Chapter 1
A 6-inch f/5 Telescope

Steven Lee

This project is typical of those suitable for someone who has a collection of leftovers from previous projects. Steven Lee raided his junk box to construct this very portable 6 in f/5 instrument for an additional outlay of about A$30. He has designed a slide focuser in order to minimise the secondary obstruction. As with all shoestring projects, the main skill required is that of finding and adapting the components – the construction itself is simple.

A low-power, wide-field telescope provides spectacular views of the heavens – the Milky Way, a bright comet or an eclipse are perfect targets. Such a telescope is an ideal second telescope to complement the high-power views of a larger instrument. It is also the right size for the junior astronomers of the family, or just for taking on holidays when you don't have room for a larger one. My telescope was made quite quickly and almost entirely from spare bits and pieces – its total cost was about $30. This is mainly because I have a large and well-filled junk box brought about by many years of telescope making – most people couldn't build it quite this cheaply. I tried to make it as simple as possible and yet be innovative in its design where I thought I could improve on standard parts.

Optics

Most people wrongly attribute wide fields of view in a telescope to having a fast focal ratio. This is not really true, but does work if you take the simple example of the same eyepiece used in telescopes of the same diameter but different focal ratios. To achieve the same wide field in a telescope of higher focal ratio, you just use a longer focal length eyepiece. In practice, the faster the focal ratio the worse the aberrations, and the better the eyepiece must be in order to cope with the faster beam; this is why the view through a slower focal ratio telescope is usually better than with a fast one. The traditional RFT (richest field telescope) is a 6 in (150 mm) f/4, but an f/5 mirror yields considerably better images for little extra inconvenience. The real benefit in the slightly slower focal ratio is not so much the lower aberrations of the mirror, but the improved performance of the eyepiece. The tube does have to be one mirror diameter longer in an f/5, but the improved image quality is well worth it. Many years ago I made a 6 in f/4 as my second telescope and its performance was fine, but this f/5 configuration is definitely better. With the same eyepiece the field of view is slightly smaller (1.7° versus 2.1°) but the quality of the field is noticeably improved. The tube is no more awkward to use and I can see no reason to use the faster focal ratio, especially with such a small telescope. In fact I would recommend never making any telescope for visual use faster than f/5 for the above reasons.

How to Calculate the Field of View for any Telescope and Eyepiece

Given f_{tel} – the telescope focal length,
$f_{eyepiece}$ – the focal length of the eyepiece, and
$fov_{eyepiece}$ – the apparent field-of-view of the eyepiece (usually stated by the manufacturer):
First calculate the magnification of the combination by:
magnification $= f_{tel} \div f_{eyepiece}$
and then the desired field:
true field-of-view $= fov_{eyepiece} \div$ magnification

As an example, a 6 in f/8 telescope has a focal length of approximately 1200 mm, while a 6 in f/5 telescope is 760 mm. A 25 mm eyepiece used in each telescope will yield magnifications of 48× and 30× respectively. If that eyepiece has an apparent field of 50°, then it will give a 1° field in the f/8 and 1.6° in the f/5. A 40 mm eyepiece in the f/8 telescope will give 30× and a 1.6° field.

I made the 6 in f/5 paraboloidal mirror on the obverse of a standard Pyrex blank which had just been used as the tool to make a friend's mirror. I used a piece of 12 mm ($\frac{1}{2}$ in) plywood as my tool, cut round with a jigsaw and coated with varnish to seal it against warping, then covered in ceramic tiles for the working surface. The mirror and tool were then ground together using standard techniques. It was polished on polishing pads stuck to the tiles and finally figured on a pitch lap on the same piece of plywood (once the tiles had been removed). I had to rush to make this mirror because I could get it aluminised (free!) if it was ready by a particular time – which was less than a week after I decided to make it. I spent several days grinding the curve, while polishing it took another 2. Figuring lasted approximately 5 minutes (in two sessions) which brought the surface accuracy to about $\frac{1}{4}$ wave – good enough for the low powers intended for this telescope.

The secondary mirror was one from my junk box, left over from a long-forgotten project. It has a small chip on one edge, but it doesn't matter for this telescope as it is well out of the on-axis field. It is a 38 mm ($1\frac{1}{2}$ in) minor-axis mirror – larger than is necessary – but as I had it on hand, I used it. A 34 mm ($1\frac{1}{3}$ in) one is the ideal mirror for this instrument.

The Tube

Fibreglass tubes are strong yet light, and totally maintenance-free – the perfect combination for a good telescope. My tube is home-made, which saves cost at the expense of a messy and smelly few days. It was originally built for the f/4 mirror I made long ago (*c.* 1973). That mirror was sold when I moved (something I regretted and was the reason for making this telescope), although I still had the tube. Of course I had to extend it because of the longer focal length mirror, but the technique came back easily to hand despite the intervening years. I had the necessary materials on hand for another project and was using this as practice. It took only a few hours of sanding, filling and more sanding to add the extra 6 in (15 cm) and smooth out the join, although it was necessarily spread over a few days. One of the advantages of this type of tube is that the colour permeates the whole job and so scratches are never seen – a boon if the telescope is mistreated or suffers when in transit. However, I couldn't match the colouring that had originally been used and so I was forced to simply paint the outside of the tube, which I did in a dark blue.

Any telescope tube should extend sufficiently far in front of the eyepiece holder to stop stray light from getting directly to the eyepiece, and to shade the area opposite the eyepiece from direct illumination. This is a failing in many telescopes and results in lower contrast images because of the extraneous light flooding the focus. The tube I made is 18 cm (7 in) inside diameter and 85 cm ($33\frac{1}{2}$ in) long, giving good shielding for the eyepiece. Finally, the tube is lined with black flock paper to make it really non-reflective. This produces a far darker finish compared with the more usual coating of black paint, with any internal reflections absorbed in the fibres of the material. Black velvet is even better, but would have cost more than I paid for the whole telescope! Besides, I had some left over from other projects and this project was intended to use up leftover bits.

The Focuser and Secondary Holder

Getting the focuser right is very important in small telescopes – not only must it satisfy all the usual requirements for a focuser (strong, light and smooth movement), but it needs to have a very low profile in order to minimise the size of the diagonal required to illuminate the field. I have always believed that a lateral-sliding focuser is the best way to achieve this, but I'd never made one, believing that they required exacting machining in order to work properly. After a lot of thought I constructed one that didn't require any machining. While it isn't perfect, it works well enough and is made almost entirely from scrap parts.

I can adjust the position of the focal surface relative to the tube over a 40 mm range, from being level to the tube surface to 40 mm above. This is a perfectly adequate range for visual use and all my eyepieces come into focus somewhere within these extremes. Because the eyepiece I intend to use most on this telescope – an old 20 mm Erfle – comes to focus with the focal surface close to the tube, it needs only a 34 mm minor-axis mirror (22% obstruction) to yield a 12 mm ($\frac{1}{2}$ in) fully illuminated field (almost 1°) and only 0.2 magnitude loss at the edge of the field. This is excellent performance and very difficult to achieve with normal up-down focusers.

The heart of a slide focuser is the slide. Instead of precision rails and ball bearings, mine is made from the discarded rails of a computer printout binder. In the old days of computers, you used to file printouts of programs (on paper 15 in (38 cm) wide) into special binders. They had cardboard covers (later plastic – just like Kydex – good for top ends) and plastic spikes which went through the end sprocket holes of the paper to restrain them. The spikes were tucked under little bits of metal which slid on metal rails. (If you don't know what I'm describing, you'll just have to take my word that these things were extremely common around computers 10 years or more ago.) Anyway, I had already cannibalised the covers of these binders for the top end of my ball-scope and I was looking at the rails wondering if they should be thrown out or put in my ever-growing junk box when I thought "rails ... slide-focuser ... hmm". And here it is (Figures 1.1 and 1.2, *overleaf*).

The sliders and rails are anodised steel, so are well protected and strong. The rails were cut to 17 cm ($6\frac{1}{2}$ in) in length, and they are kept at the right distance apart by two cross-members of 10 mm ($\frac{3}{8}$ in) wide, 2 mm ($\frac{1}{16}$ in) aluminium. Holes through these two pieces are used to bolt it to the tube. Longitudinal pieces of 1 mm ($\frac{1}{32}$ in) aluminium bent into a right angle add extra stability and allow a mounting point for the driving mechanism. The moving part is a plate of 2 mm thick aluminium approximately 70 mm ($2\frac{3}{4}$ in) wide and 100 mm (4 in) long. A $1\frac{1}{4}$ in (31.8 mm) hole is at one end of the plate, with an aluminium tube over it to hold the eyepieces. Normally, such a tube would need to be machined, but I had one in my junk box from a previous project. Motion is provided by a rack-and-pinion drive from (you guessed it) my junk box. My father made this for me for my very first telescope – a $4\frac{1}{4}$ in (108 mm) f/12 Newtonian – which was decommissioned some years ago. The pinion gear rides on a chrome-plated steel shaft removed from a floppy disc drive, which in turn rides in brass blocks which simply have suitable holes drilled in them. The hand knob is from another

Figure 1.1 Slide focuser.

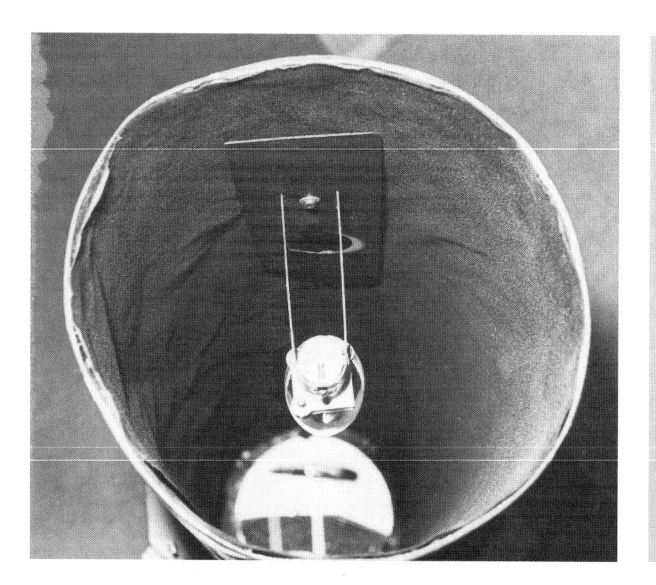

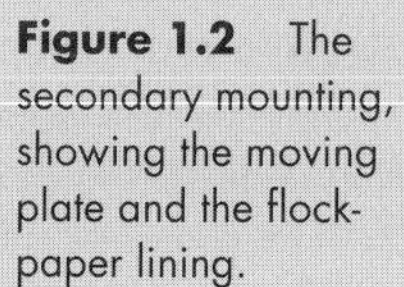

Figure 1.2 The secondary mounting, showing the moving plate and the flock-paper lining.

focuser which was removed when fitting a motor drive and placed in a junk box to await re-use. (The shaft end of the knob was used as a coupling between the focuser and encoder on my 20 cm (8 in) f/4.5 Newtonian imaging telescope – nothing wasted here!) There is one other important feature in this mechanism which makes it a delight to use – a 6:1 reducer. This is a commercial part, sold to amateur radio builders as a reducer for a radio tuning knob; current price is of the order A$20 (but I had one in my junk box from a previous abandoned project). It is gearless, the reduction being done by friction-coupled balls turning between the input and output shafts. The addition of this mechanism turns the focuser from an ordinary one into an exceptionally nice one.

The secondary mirror must be attached to the moving plate and positioned so that its centre (optical, rather than geometrical) is directly under the eyepiece holder. I use a U-shaped piece of 1 mm thick aluminium about 20 mm ($\frac{3}{4}$ in) wide to link the plate to the secondary holder. A suitably sized block of aluminium on the plate keeps it at the right spacing, while the bottom of the U wraps around a 20 mm-diameter aluminium tube (a cut-off portion of one of the truss tubes on the ball-scope in Chapter 8). A single screw and nut holds the two together and allows for rotation of the secondary should it be necessary. Slots in the top of the U allow for positioning the secondary mirror accurately under the eyepiece. Mounted on the same block on the plate is a piece of plastic used to shield the secondary from light getting through the slot in the tube in which the focuser slides.

Getting back to the secondary holder, the end of the tube has a small, flat 2 mm aluminium plate glued to it through which three collimation screws are located. The secondary mirror is attached with silicon sealant to another 20 mm tube cut at 45°. The other end of this tube also has an aluminium plate glued to it through which the other end of the collimation bolts go. The spring-tensioned collimation bolts are arranged not at the "standard" 120° spacing, but rather so that adjustments occur at right angles (up-down and left-right as seen through the focuser). One screw acts as a pivot and is only touched if the whole assembly needs to be moved towards or away from the primary mirror; only the other two are used when collimating.

The Primary Mirror Cell

The primary mirror cell (Figure 1.3, *overleaf*) is one 4 mm ($\frac{3}{16}$ in) thick aluminium plate, the mirror resting

Figure 1.3 Primary cell with the collimating bolts in a right-angle configuration for easier collimation.

on a three-point support and held laterally by three posts attached to the plate. Through each post is a bolt (with locknut) to securely position the mirror. The collimation bolts are on the outside of the tube and move the plate relative to the tube. Figure 1.4 shows two of the bolts and the attachment points on the outside of the tube. They are easily accessible while looking through the eyepiece, a boon for easy collimation. I made the primary cell adjustments operate in the same way as I did for the secondary mirror cell. Up-down, left-right adjustments are far superior to the traditional triaxial method and I don't understand why people still insist on making them.

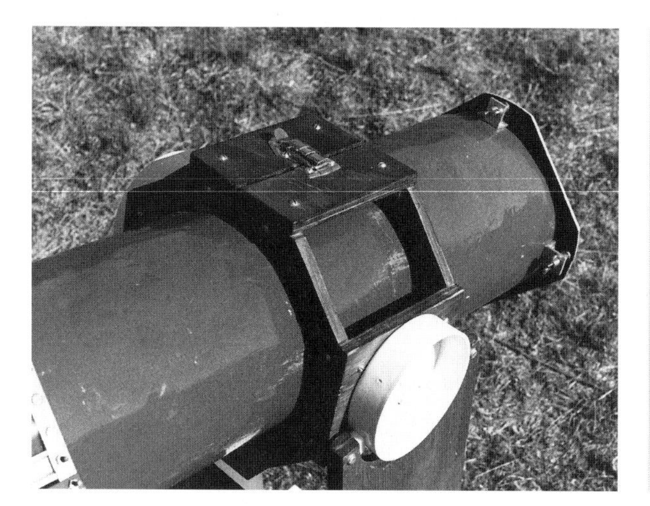

Figure 1.4 Mounting cradle and altitude trunnions. The primary cell and two of its collimation bolts are shown on the right-hand end of the tube.

The Mounting

A simple altazimuth mounting is ideal for this type of telescope. This was one mistake I made in my earlier version which had an undriven equatorial mount. I quickly discovered that an equatorial without a motor is a pain – it still has to be pushed about, but the eyepiece always seems to be in the wrong position for comfortable viewing. I intended that this telescope would be better, and designed it to be used with the observer seated (which I consider the optimum way to observe) and to have a rotatable tube for maximum comfort.

The mounting represents most of the cost of this instrument, as I had to buy most of the parts for it – there were not enough large pieces of timber in the junk box this time. As can be seen from the accompanying photographs, the tube is held in a cradle that can be quickly opened to allow the tube to be rotated to bring the eyepiece to the most comfortable position for observing. It also allows the tube to be repositioned to accommodate eyepieces of different weights. This was the hardest part to make, taking me almost a whole day to cut out and assemble; not that it was difficult – it was just fiddly. It is made from $\frac{1}{2}$ in (12 mm) plywood cut out with a jigsaw. The outside is octagonal, the two rings being separated by 15 cm (6 in) and held by four plywood plates. It was assembled with the top, bottom and side pieces whole before cutting the top and bottom pieces for the hinges and latch. Because plywood can't be end-screwed, all the pieces are held together by small aluminium angle and small screws and nuts. This is where all the time went – there are 50 screws holding everything together, which meant that almost 100 separate holes had to be drilled. But it was all worth it in the end, as it is a delight to use because it is so easy to adjust.

The rest of the mounting is pretty standard Dobsonian technology. The altitude bearings are 12 cm in diameter and are end-caps for sewer pipe and cost A$4 each. They ride on Teflon blocks attached to the mounting. Because the tube assembly is so light, the Teflon blocks had to be moved far apart to increase friction, or else the tube moved far too easily.

The height of the mounting was set by the height of the eyepiece when at the zenith and with the observer seated in a chair. (If one did make it as low as could be done then it would be very uncomfortable to use, because the eyepiece would be so close to the ground. It would require a table or some other means to raise it high enough – although unraised it might prove to be ideal for children's use.) The height to the centre of the altitude bearings is a bit under 70 cm. The side boards are leftover pieces of Craftwood (a brand name for medium-density fibreboard – MDF – a dense, reconstituted timber product) screwed to a piece of chipboard which was covered in Formica (left over when our kitchen was built). The front board is another piece of $\frac{1}{2}$ in plywood. Note the handle cut into its face to facilitate easy carrying. The usual three pieces of Teflon attached to a plywood base finish it off. All the wood was stained before three coats of varnish (I had to buy a new pot of varnish – more expense!) was applied. The completed instrument is shown in Figure 1.5.

Figure 1.5 The completed telescope.

Conclusion

While I was able to make this telescope for very little actual cost, I relied heavily on my large collection of spare parts – the real cost of the telescope is somewhat higher. Depending on whether the optics are made or purchased, and how capable you are at bending metal into the desired shape, it may cost several hundred dollars to make. But remember that it doesn't need to be perfect as it is only intended for low powers.

In use, the telescope is an absolute joy. When viewing from the chair one tends to stay in one field for a while before moving the chair and telescope around to another region. Depending on your observing style, you can either consider this a problem or perfect observing. You'll notice that there isn't a finder on the telescope. I don't think that it needs one, because I just sight up the side of the tube and then look through the eyepiece. If the object in question isn't in the field, then it will be close enough to find without a long search. (Remember, you don't go searching for 18th-magnitude galaxies with this telescope. It is for low-power views of bright objects.) Some readers might think that one of the unit-power finders is called for with this telescope, and I may make one for it one day.

With the usual 20 mm Erfle it yields 38× and a field of about 1.7° – perfect for comets, eclipses and just wandering through the Milky Way (the Carina region is superb, with nebulosity filling the whole field). However, a 16 mm Nagler also yields a field of 1.7° (at 47×) and better images at the edge of the field, and so has become my eyepiece of choice with this instrument. If I had made the focuser 2 in instead of just $1\frac{1}{4}$ in, then I could have used a 20 mm Nagler for a 2° field, or my 32 mm Widefield for 24× and 2.4°! It is tempting to rebuild the focuser for this last combination, because such wide-field views are truly stunning and are what RFT viewing is all about.

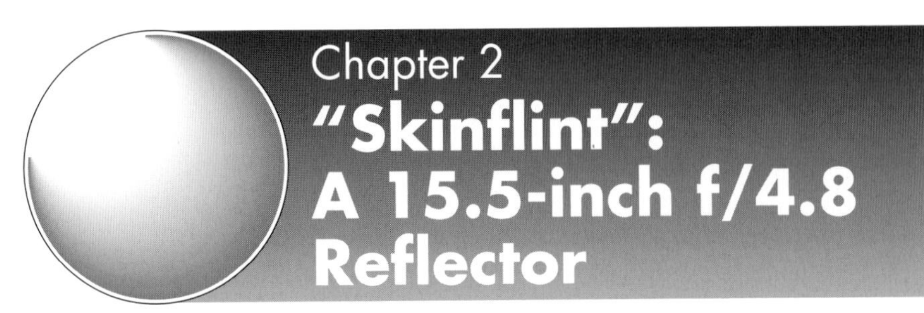

Chapter 2 "Skinflint": A 15.5-inch f/4.8 Reflector

Gil Stacy

A useful talent that the maker of the shoestring telescope to acquire is that of being able to adapt items which are not normally associated with telescopes or optics. Gil Stacy has found creative uses for commonly available materials and artefacts, much of which many people would class as scrap, to build his "Operation Skinflint", a 15.5 in f/4.8 Dobsonian reflector for a cost of less than $400.

"Cosmic Voyager, Destination Unknown". Thus reads the inscription on the grave of Conrad Aiken, the former Poet Laureate of the United States, who lies buried beneath the Spanish-moss-draped oaks of Bonaventure Cemetery in my home town of Savannah, Georgia. Aiken had noticed the words when they appeared in the local newspaper's arrivals and departures of ships into and out of the coastal port of Savannah in the south-eastern US. Not far from where Aiken lived and died, local shipbuilders launched 88 Liberty Ships during World War II. These vessels sailed the wartime North Atlantic, supplying the Allied effort. Another cosmic voyager, my telescope, was born a surplus part of one of these heroic ships, which itself had been born in a shipyard on the banks of the Savannah River six decades ago.

In 1988, Henry Smith, a friend and a former ATM, gave me an assortment of plate-glass portholes. Henry had collected portholes from the old shipyard area of the port of Savannah. Used at first as windows to the seven seas, 60 years later one of the 15.5 in (39.5 cm) portholes has become a window to the cosmos, travelling from galaxy to galaxy under darkened skies.

Figure 2.1 The author with "Skinflint".

The cradle of the modern ATM movement is in Springfield, Vermont, where out-of-work machinists ingeniously crafted telescopes from another's junk. Beginning in the sixties, John Dobson, a San Francisco Vedantic monk, furthered the ATM tradition by breaking taboos of mirror thickness ratios and eschewing machined metal. From paper and wood, he built portable large-aperture telescopes using scraps found at construction sites. With my roots firmly planted in those movements, I decided to make a quality telescope as cheaply as possible from low-cost or no-cost materials, found either in scrap heaps or at bargain prices. According to my ever patient wife, Louisa, it also appears to be made from missing household tools and objects, including a broom handle. The completed project(Figure 2.1), a 15.5 in aperture f/4.8 in a two-spar Dobsonian telescope, costing $394.00, bears the name "Skinflint".

The Mirror

The porthole is 15.5 in in diameter and 1 in (25 mm) thick. The finished weight is a scant 11 lb (5 kg). Grinding, polishing and figuring was an "on again, off again" project, during which I also made numerous smaller mirrors and lenses. Using techniques popularised by John Dobson and Bob Kestner, the mirror was ground by hand technique, with the rough grinding done outside in my backyard, with fine grinding, polishing, and figuring performed in my kitchen. To the "delight" of my family, testing was accomplished in my kitchen and dining room by traditional Foucault and Ross Null techniques.

Rough grinding was a chore; taking the porthole from flat to a sagitta of approximately 0.2 in (5 mm) took 20 hours of actual grinding time, not including charging with abrasive and cleaning between wets. The grinding tool I used was a 16 in (41 cm) concrete patio or garden stone, cast as a disc in concrete, and purchased for $3.00 at a local hardware store.

I sealed the disc with high-quality epoxy left over from a boat project. After it was dry to the touch, I epoxied 2 in (5 cm) square ceramic tiles to the surface of the stone. Since the target sagitta and tile thickness were both 0.2 in (5 mm), the tiles eventually became too thin at the edges and I had to secure fresh tiles across the total surface.

The cost of grinding and polishing was fairly low because I split with three other amateurs a bulk purchase of grinding and polishing materials. I polished with paper pads and then finished with pitch.

An acknowledgement of gratitude is owed to the late Tom Waineo, a master optician, who assisted in the final figuring process with his practical suggestions. I had run into a brick wall of frustration, unable to eliminate a troublesome zone, until helped by Tom.

I had read accounts of parabolising a large mirror, with one account stating the author's ease in figuring with a 50% subdiameter lap. Using the same technique, my efforts failed. If one visualises the graph line of a perfect paraboloid as being flat, my problem was as follows: Moving from the centre of the mirror to the edge, the graph line climbed from the centre, peaked at the 70% radius zone (a radius of 5.5 in from the centre), then descended to the edge. In other words, at the 70% radius, the mirror's surface was too high in relation to the centre and edge zones. Because I had no one available locally to turn to for guidance, the project went into limbo for 2 years. It was then that I discovered the telescope making forum on CompuServe. A familiar name appeared regularly, giving advice to all who asked. I recognised his name, Tom Waineo, as that of a contributor over the years to *Sky & Telescope* magazine. Tom had retired from a career as a professional optician, and was spending his time helping amateurs with advice and making custom mirrors. Both Tom and I used Dick Suiter's Foucault reduction program ADMIRR. With my Foucault readings sent via Email, Tom could instantly see the same profile that I was seeing, 500 miles away. By Email, Tom advised me to make a 2×6 in rectangular lap which would aggressively lower the 70% zone's peak without further lowering the centre or edge. The brick wall quickly fell and I was quite pleased with the smoothness of the figure and its overall profile.

Bob Fies of San Carlos, California, coated the mirror. He has a small operation in his garage and accommodates amateurs, including John Dobson, in the coating of mirrors for a competitive price. The January 1997 price for coating my mirror was $50.00. I purchased a high-quality 2.5 in (63 mm) diagonal from E&W Optics for $75.00. The price had been reduced because of a tiny sleek on the surface.

The Mirror Box

I constructed the mirror box (Figure 2.2) from $\frac{1}{2}$ in birch plywood in traditional Dobsonian fashion, with a tailgate connected to the bottom via a piano hinge. I nailed and glued the sides together with a $\frac{3}{4}$ in (18 mm) pine strip in each corner for added rigidity.

Self-tapping drywall screws secure the tailgate's closure. A sling of military surplus nylon webbing supports the mirror.

Rather than using the tailgate, I find it convenient to place the mirror in its support system through the box's open end rather than by opening the hinged tailgate.

David Chandler's computer program calculated the 18-point suspension system. As suggested by David Kreige's article in issue 35 of the no-longer-published magazine *Telescope Making*, I mounted one end of the sling through a threaded rod split with a hacksaw blade. From the outside of the mirror box, the split threaded rod's rotation controls sling length. A nut on the inside of the box helps to lock the rod in place.

The eighteen-point suspension system utilises six triangles cut from 0.25 in plywood. The "points" are the vinyl-covered heads of thumbtacks. I cut the

Figure 2.2 The primary mirror cell, showing the 18-point suspension and the broomstick sections mentioned in the text.

"teeter" or balance bars from aluminium scrap. Each "teeter" bar balances on a fulcrum, which is nothing more than a nut threaded on a carriage bolt. I hacksawed the round "button" from the end of the carriage bolt, leaving the square base attached to the bolt which prevents the balance bar from slipping away from the bolt assembly. The "button" stood above the suspension points, necessitating its removal. Blue Loctite secures the fulcrum nut permanently on the carriage bolt. The bolt in turn threads through the bare wood of the tailgate assembly, with the three carriage bolts spaced 120°. On the outside of the tailgate, each carriage bolt controls mirror tilt adjustments. On the end of each bolt, I secured with Loctite a plywood knob and T-nut assembly.

With the mirror in the sling, four pieces of broomstick prevent lateral movement. This technique was first described by David Kreige in *Telescope Making* no. 35. Each dowel has an off-centre hole bored end to end which allows lateral adjustment of each dowel in order to barely clear the mirror's edge. A nut and bolt secure the dowel to the tailgate. The mirror does not touch the dowels. Four mirror clips are fastened on top of each dowel. The clips are eyebolts with the threads covered with short sections of PVC tubing. The dowels' through-bolts secure the eye of the eyebolt to the top of each dowel. With the mirror in place, I rotate the ends of the eyebolts across the top of the mirror. The eye-bolts barely clear the mirror's surface.

Inspired by Steven Overholt's wonderful book *Lightweight Giants*, I made the mirror box's oversized altitude bearings from 8 mm Styrofoam 1.5 in (38 mm) thick sandwiched between two pieces of door skin plywood, 0.125 in (3 mm) thick. I salvaged the Styrofoam from a trash heap. On each of the bearings, I nailed and glued the door skin to an 8 mm strip of pine 1.5 in (38 mm) wide which sits flush with the straight edge of each bearing. A coping saw blade trimmed the excess Styrofoam after the door skin was glued to the Styrofoam with common carpenter's glue. Inexpensive fibreglass-reinforced plastic (FRP), pebbly side outward, surfaces the curved portion of the bearings. To prepare the FRP's surface for the carpenter's glue, contact cement was first applied to the FRP. Carpenter's glue and drywall screws secure each end of the FRP strip to pine strip inside each bearing's straight edge. Water-based polyurethane varnish covers the exposed wood of the bearings, and three through-bolts per bearing anchor the bearings to the mirror box.

The Spar System

For simplicity, I adopted a two-tube spar system after seeing an article in *Amateur Astronomy* no. 7 which featured Ron Ravneberg's beautifully crafted 8 in (20 cm) travel scope, named "Alice". Ravneberg made the scope for the trip of a lifetime to the Australian outback. While others, such as Thane Bopp, had previously made two-spar systems, Ron Ravneberg's "Alice" raised the design to an art form. For spars, I obtained two aluminium pipes 0.125 in (3 mm) thick, of 2 in (50 mm) OD. Each weighs 6.5 lb (3 kg). The focuser board, a piece of 0.5 in (12 mm) birch plywood, sits on the outside of the aluminium pipes. Along the outer edges of the focuser board are parallel pieces of aluminium channel screwed to the board with the opening of the channel facing the pipes. Each pipe nestles against the channel's two edges, and is drawn tight with a through-bolt, tightened with an plastic knurled knob purchased from the Reid Tool Company. The system allows quick takedown and reassembly with repeatable, accurate alignment.

The pipes anchor full length inside the mirror box against a wooden channel, the latter formed by the box corner's 0.75 in × 0.75 in (18 mm × 18 mm) pine strip, and a parallel strip, separated from the corner brace by 1.5 in (38 mm) and attached with screws. Two knurled knobs per pipe attach the spars to the mirror box, and pull the tubes tightly into the channel, giving the same repeatable, accurate alignment of the focuser assembly.

The Focuser Board Assembly

The focuser board assembly (Figure 2.3)consists of a 0.5 in (12 mm) thickness of birch plywood, aluminium channels to secure the spars, the focuser, spider assembly and a low-priced air rifle 'red dot' sight used as a

Figure 2.3 The focuser board assembly and diagonal holder.

unit power finder. In order to place the focuser as close as possible to the diagonal, a circle was cut from the centre of the board and a piece of door skin was glued and nailed to the diagonal side of the focuser board, after which the focuser was bolted to the 0.125 in (3 mm) plywood.

The Diagonal Holder

A 6 in scrap section of the spar tube is the diagonal's stalk holder. A 1 in diameter section of broomstick, approximately 8 in (20 cm) long, serves as a stalk to which the diagonal is attached. One end of the stalk was cut to a 45° surface. A piece of 0.25 in (6 mm) plywood, cut to the diagonal's outline, attaches with a screw and epoxy to the diagonally cut stalk. I drilled holes 120° apart inside the edges of the plywood, which has a polyurethane varnish coating. I squeezed silicone glue into the holes, and used pennies as spacers, and with a vice levelling the plywood, I lowered the diagonal, back side down, onto the silicone blobs. After a 24-hour curing period, I completed the installation of the diagonal and two-vane spider assembly.

Along the interior side of the spar channels, I attached a section of 1.5 in (38 mm) aluminium angle to the focuser board on opposite sides to allow attachment of the two-vane spider assembly. Using a scrap aluminium strip, I cut and bent the 0.1 in (2 mm) thick and 1.5 in (38 mm) wide strip to shape. Strips of wood clamped to the strip facilitated controlled and defined bending.

Prior to attaching the diagonal holder tube to the vanes, I drilled and tapped four sets of opposing holes into the stalk-holder tube near each open end to accommodate recessed hexagonal-headed machine screws of $\frac{1}{4}$ in × 20 tpi. ($\frac{1}{4}$ in UNC; M6). These screws hold and provide longitudinal, lateral, and tilt adjustment of the diagonal stalk. The diagonal holder tube attaches to the vanes with two machine screws. It became apparent that the aluminium tube's softer threads were losing ground to the steel screws. The screws would eventually loosen, requiring more adjustment than I felt was necessary. Loctite did not hold because the aluminium holes were becoming oversized. The problem was solved with four Teflon strips, 0.125 in (3 mm) thick, 0.75 in (20 mm) wide and 5 in

(125 mm) long. I drilled two holes per strip to accommodate two screws per strip. I inserted each Teflon strip inside the tube lengthways; the top and bottom screws were tightened through each strip, with the screw tips gripping the stalk. Because of the nature of the Teflon, the holes in the strips tend to close up, rather than loosen and tighten around the threads, locking the screws in position.

Since the Teflon strips are secured at each end by the four pairs of screws, the strips do not require anchoring to the wall of the tube in order to hold the screws tightly. After installation into the holder tube, the eight screws adjust the stalk's position. The screws, while held firmly by the expanding Teflon, easily adjust with an Allen key. To prevent the unthinkable, a piece of coathanger wire passing through a hole in the sky end of the stalk serves as a safety-catch.

Since the focuser faces open sky, baffling is needed to prevent unwanted light flooding into the eyepiece. After eliminating other methods by trial and error, I settled on surrounding the diagonal with a 5 in (125 mm) diameter section of aluminium thin-walled tubing left over from a refractor project. On wrapping paper, I drew a full-scale mock-up of the light cone returning to the eyepiece. I then cut the tubing to a length that would not infringe on the return light cone, and a hole in the tube's side allows the return light cone to enter the focuser. Matt-black spray paint covers the entire assembly.

When the scope is not in use, the focuser board assembly fits into slots made by screwing wood strips into the sides of mirror box. A door skin lid seals the box from dust and unwanted hands. Screen door handles attached to the box sides permit easy handling and transport. A large corrugated cardboard pizza box, reinforced with foam board and fibreglass tape, protects and stores the mirror when not in use.

Ground Board and Rocker Assembly

I constructed the rocker assembly (Figure 2.4, *overleaf*) using $\frac{1}{2}$ in thick birch plywood, $\frac{1}{8}$ in thick door skin and foam core. While others had written articles regarding composite lightweight construction of telescope

Figure 2.4 The mirror box and rocker assembly.

materials, Steven Overholt's book, *Lightweight Giants*, detailed a "low-tech" approach to composite manufacture utilising thin wooden door skin, carpenter's glue and Styrofoam building panels. I made Skinflint's sideboards from a frame of 1.5 in × 0.75 in (38 mm × 18 mm) pine boards with a panel of 0.75 in (18 mm) thick Styrofoam inserted into the frame and the entire board covered with door skin glued and nailed to the frame. Placing the two sideboards on a flat surface, I stacked weight on each completed sideboard during the overnight drying period. The bottom of the rocker and front board are $\frac{1}{2}$ in (12 mm) birch plywood. The top of each side board has a curved cut conforming to the curve of the side bearings. The pine board frame was deep enough to prevent cutting into the foam panel. Two pieces of Teflon per each curved section allow the side bearings to pivot smoothly in altitude. A piece of FRP covers the bottom of the rocker assembly. This in turn sits on the ground board, held by a through-bolt.

The ground board is a triangular structure of door skin glued to 1.5 in (38 mm) thick foam surrounded by a pine frame of equal height. Three hard rubber furniture castors are mounted on the points beneath the board. The rocker assembly pivots on three equally spaced pieces of Teflon attached to the ground board.

Counterbalancing

After construction, the scope seemed to be too wobbly, and vibrations were slow to dampen. The focuser side of the scope weighed at least 13 lb (6 kg) more than the opposite side. A conventional eight-tube truss system distributes the weight more evenly around the optical axis. The extra 13 lb on only one side of the optical tube assembly of Skinflint created the instability. Counterweighting the side opposite the focuser board with lead shot in PVC tubing eliminated the problem. The weight hangs by a wire hook on the open end of the mirror box.

Assembly and Use

Before an observing session, to get a head start on equilibration, I leave the mirror on the front porch in its pizza box carrying case. Order of assembly is no different from that for any other truss system. I install the mirror only after the focuser board is in place.

The telescope takes less than 5 minutes to set up and use (Figure 2.5, *overleaf*). Even with the mirror stored separately from the mirror box, collimation with a Cheshire eyepiece takes but a few minutes as the collimation dot in the middle of the mirror usually appears close to centred in the bull's-eye of the Cheshire. With my 32 mm Widefield Nagler, the scope loafs along at 60× magnification, with a crisp field of view of 1.1°. Slight astigmatism in the mirror's surface limits the mirror from being first-rate, with the astigmatism being noticeable at 200× magnification. While definitely not a planetary scope with its present limitation of astigmatism, it has four times the light-gathering ability of my 8 in (20 cm) f/5 Newtonian, and springtime cruising with my 32° TeleVue Widefield through the galaxy clusters of Coma Berenices and

Figure 2.5 "Skinflint" unassembled for transport and assembled.

Virgo is a stunning visual journey. Crisp star points fill the entire field of view in the Widefield. With Skinflint, I have bright, clear views of nebulae, galaxies and other deep sky objects that are faint images at best in my smaller scopes. For dim, deep sky objects, nothing beats aperture except more aperture, and even with slight astigmatism, the views are brighter and more spectacular than in my finely figured 8 in scope.

Lessons Learned

Rough Grinding

Had I known of Bratislav Curcic's steel ring technique of rough grinding, the process would have been speedier and I would not have had to replace tiles ground thin during rough grinding.

Astigmatism

I had tested the mirror for astigmatism with a variety of tests, but apparently not well enough and misinterpreted a test which disclosed it. A simple star test would have alerted me to the problem. On other projects, I had star-tested the optics during polishing and figuring. In those earlier projects, the optical tube assembly was completed in advance before figuring. However, a bad case of laziness prevented me from making the optical tube assembly for this project in advance. Astigmatism is something that I have avoided in the past with other projects, but I had never attempted a thin, large mirror until this project. Astigmatism is avoidable in thin mirrors and I will redouble my efforts to eliminate it in the future.

In hindsight, I believe the astigmatism was detected in the Ross Null test used in the Ronchi mode. Before beginning Skinflint, I had made a 4 in (10 cm) Ross Null lens from Schott Grade A BK-7. I had difficulty in setting the test up in the Ronchi mode. The lines would spread out slightly at the bottom. I was able to set up the test with a slit and had no trouble using it to null the mirror. However, rather than blaming the mirror for my perceived difficulty in Ross Null Ronchi testing, I foolishly blamed the test for being inherently difficult to set up. I now realise the "difficult" Ronchi pattern indicated astigmatism. It is said that pride usually goes before the fall, but in my case, it lasted until the coated mirror was returned.

Confident I can do better, I am at present making another mirror, a 16.5 in (42 cm) of similar focal length, that I will complete before temporarily "blinding" the present mirror when I refigure it. With a simple modification of the optical tube assembly, I will easily be able to adapt the scope to accommodate the second mirror. The optical tube assembly was built to accommodate mirrors up to 18 in (46 cm) in diameter. I do not foresee any great difficulty in eliminating the astigmatism in the first mirror, as it probably resulted from inadequately flattening the back of the mirror before I began grinding.

"Measure twice, cut once." In my case, perhaps I should measure four times in order to cut or drill once.

Study the picture of the focuser board and you can see the need for the maxim.

Spar Assembly and Counterweighting

The 2 in, 0.125 in (3 mm) thick aluminium tubes are probably overkill. Thinner tubing would suffice and without much doubt eliminate the need for counterbalancing. Rather than two knobs, one knob per lower spar tube is likely to be sufficient to firmly lock each tube into the mirror box.

All things considered, the project is a success. I have built a usable, easily transportable telescope of large aperture for under $400.00. Had I made my own focuser, an additional $130.00 could have been saved from the project. In addition to owning a real-time window to the deep sky, the project generated a feeling of accomplishment. There is also a sense of satisfaction from using simple tools and economical materials to make an instrument capable of viewing reaches of the universe as they appeared millions of years ago.

The educational process of making this mirror has fuelled a desire for bigger and better. Judging from the reaction of my friends and family as they stare at deep sky wonders through the eyepiece, Skinflint is an unqualified success. Perhaps others will be inspired by the success of the contributors to this book to craft their own cosmic voyagers for destinations unknown.

Summary of Costs in 1997 US Dollars

Mirror – a 15.5 in × 1 in porthole given to me:	0.00
Tile tool: 2 in ceramics on a 16 in patio stone (used leftover epoxy):	8.00
1.5 kg Gulgoz 73	15.00
Grit and polishing compound. Estimate, shared bulk order, but lots left over and used in other projects:	15.00
Coating, shipping	70.00
Plywood, 0.5 in birch	40.00
Door skin, 0.125 in	7.00
Foam for rocker box, bottom board and oversized 21 in bearings. Found in Dumpster (dustbin to UKers)	0.00
2 aluminium pipes, 72 in long, 2 in OD	30.00
8 ft × 4 ft sheet of FRP (fibreglass-reinforced plastic) used in bearings (used about 4 square feet) cost for sheet was $30.00, but rest is already spoken for in other scopes	4.00
Teflon, scrounged	0.00
Focuser recycled from older scope	0.00
Focuser purchased new	130.00
Secondary, scratch and dent sale from E and W 2.5 in minor axis	75.00
Paint and polyurethane varnish, left over from home repairs	0.00
Spider made from scrap aluminium	0.00
Total (inclusive of focuser)	**$394.00**

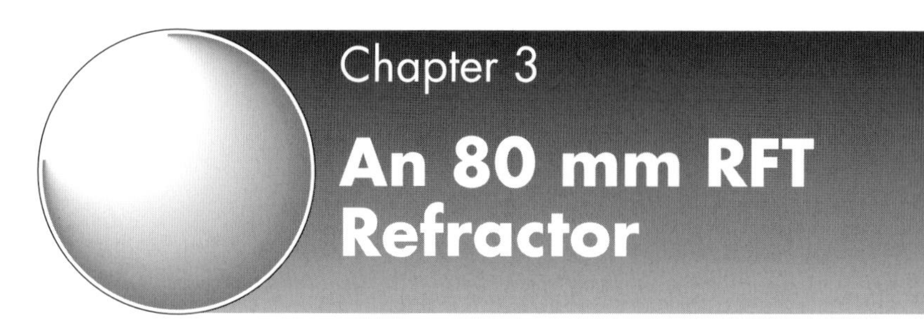

Chapter 3 An 80 mm RFT Refractor

Stephen Tonkin

This telescope is typical of many shoestring projects in that the inspiration comes from the serendipitous discovery of an important component; in this instance it is the objective lens. This initial discovery inspires creativity in the adaptation of everyday materials, and so the telescope develops.

A star-party report in the *News and Notes* of the astronomical society to which I belong made reference to a telescope that appeared to be constructed from "an old aero lens and plumbing bits". "Surprisingly," the article continued, "it gave very good images." I have to confess to experiencing similar pleasant surprise when I gave it first light.

The Objective Lens

Friends and relations well know that I am always on the lookout for anything that might be usable as part of a telescope, so I was not overly surprised when my sister-in-law telephoned to say that she had found something in a junk shop. This "something" turned out to be a reconnaissance camera lens dating from World War II. Although the front aperture appears to be a little over 100 mm (4 in) in diameter, inscribed on the heavy brass lens cell is the information that the lens has a focal length of 20 in (508 mm) and an aperture f/6.3; a quick calculation will confirm that the effective aperture is just over 80 mm ($3\frac{1}{8}$ in). The discrepancy is due to the presence of an iris which can reduce the aperture to f/16 (32 mm), a facility which later proved to be very useful. I have since obtained similar lenses from military surplus stores.

I had no idea how the lens would perform as a telescope objective, so the only option was to fit it to a tube and eyepiece and find out.

The Tube Assembly

Experimentation revealed that the lens cell is a tight fit into a piece of standard 4 in (110 mm) PVC soil pipe. Some years previously I had discovered that some canned food cans are a good fit into the small end of a pan reducer, the piece of plumbing that connects the outlet of the toilet bowl to the soil pipe, and so the (unused!) tube components were found.

Although the lens cell is a tight fit in the pipe at room temperature, I did not trust friction alone to hold things together, especially on an instrument that would be used on cold nights. I had a section of plastic pipe that *almost* fitted over the soil pipe, so I cut a ring from this and split it to make a sliding collar that could butt up to the flange on the pipe. Three L-shaped aluminium struts secure the collar to a flange on the lens cell. A nut and bolt tighten the collar onto the tube via two small pieces of aluminium angle, one of which is secured to each end of the collar.

An "own brand" baked-bean tin from the local supermarket was an excellent fit in the small end of the pan reducer. In order to make a neat hole for the focuser draw-tube, I inserted a piece of 50 mm square timber into the tin and secured the end of the tin to it with screws through small holes positioned so that they could later be used for securing the focuser to the tin. These small holes were not drilled, but made with a diamond-point awl. This timber supports the end of the tube against deformation while it is drilled. With the timber clamped in a vice, I drilled through the metal and into the wood with a hole-saw and cleaned up the edge of the hole with a small file. I then used tin-snips to cut the tin to the appropriate length. To complete the preparation of the tube parts, I roughed up the internal surfaces with some abrasive paper and gave them a coating of matt-black paint.

I had a focuser (see below) from another project. I fitted this to the tin, and fitted the tin to the pan reducer. I drilled each end of the pan reducer with three radial holes which through which I could bolt it to the tin and to the soil pipe. The holes are spaced at not quite 120° intervals. This unequal spacing is useful for ensuring that the arrangement can only be assembled in one orientation. I bolted the tin to the pan reducer, but not the pan reducer to the soil pipe.

A glance down the draw-tube revealed that there were no nasty internal reflections, mostly, I think, owing to the light-baffling nature of the internal flanges on the plumbing parts.

The Focuser

I initially made a rack-and-pinion focuser from $3\frac{1}{2}$ in (90 mm) length of $1\frac{1}{4}$ (32 mm) OD brass tubing. This does not accept the standard $1\frac{1}{4}$ in eyepieces, so I constructed a star diagonal from some $1\frac{1}{4}$ in ID brass tube salvaged from an old film projector. After cutting it in a mitre block in order to be able to refit the pieces at a right angle, I inserted a prism (from an old pair of binoculars). The prism is epoxied to the head of an M3 bolt. The bolt protrudes through a hole in the tube and, after adjusting the prism by twiddling with the bolt, I fixed the prism in place with liberal amounts of hot-melt glue. I used hot-melt because it was the only adhesive I had to hand apart from a very small amount of epoxy – I was pleasantly surprised that it did not crack the glass! At one end of this diagonal I drilled and tapped a hole to accept the knurled-head bolt which secures the eyepiece (Figure 3.1, *overleaf*).

I found a cog to serve as a pinion in my junk box. I made the rack from two brass bolts, whose thread fits the teeth of the pinion, by decapitating them and filing them flat along one side. I filed a flat onto the brass tubing and soldered the flat side of the bolts to it, butting the bolts end to end.

I modified a brass pipe-flange with file and drill in order to form the pinion housing and widened the hole in it so that the brass tube was a sliding fit . The pinion is mounted on a brass rod with a small Bakelite knob at the end. While it is unlikely that you will have the same junk in your scrap box, with a little imagination and ingenuity it is possible to concoct a reasonable focusing arrangement.

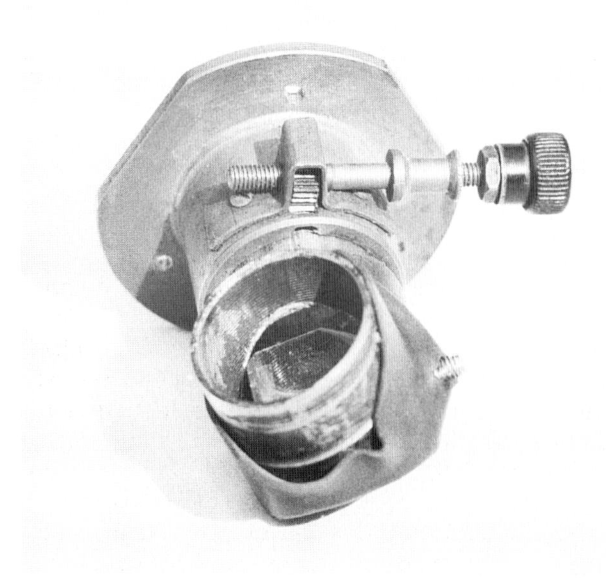

Figure 3.1 The original focuser and star diagonal. The prism is visible inside.

Collimation

My collimating tool is a plastic 35 mm film can in which I have drilled two 2 mm holes, one each in the dead centre of the base and the lid. I have found that the easiest way to collimate a refractor is to do so in two stages. First, remove the objective and tape a cross-"hair" of string across the "big end" of the tube. The film can goes into the focuser and the focuser end of the telescope is waggled until the cross is visible through the holes in the can. Then mark the soil pipe through the holes in the pan reducer and drill bolt holes in the former. Bolt these parts together and recheck the collimation.

The second stage is to square the optical axis of the objective to that of the draw-tube. If a light on the central axis of the draw-tube reflects off the objective, there will only be one reflection at the centre of the draw-tube when the objective is squared on. This may seem a bit "kludgy", but subsequent star testing suggests that it is adequate. On this instrument, no adjustment was needed but, had this not been the case, I would have filed the objective end of the soil pipe in order to obtain the correct orientation of the objective.

The Finder

An instrument of this type does not need a finder telescope – indeed, I have used it as a finder on a 12.5 in Dobsonian – and for some years I targeted it merely by sighting along it. More recently I have been experimenting with unit-power illuminated finders and I have fitted one to this telescope with very satisfying results.

The finder consists of a brass tube, a small biconvex lens, a piece of aluminium foil, and a red LED. The principle is simple: the LED illuminates the pinhole, which is at the focal plane of the lens. The lens therefore projects the image of the pinhole at infinity. The finder is used with both eyes open, one eye looking into the lens, the other looking at the sky. The observer's brain superimposes the image of one eye upon that of the other and the visual effect is that a red dot is placed on the sky.

Some experimentation will probably be necessary to obtain a suitable pinhole. The method I have found to be most satisfactory is to place the foil on a metal surface and to use a very fine needle. I rotate the needle, like a drill, between finger and thumb and use very light pressure.

I cut the foil with the pinhole into a disc slightly larger than the brass tube. It was simple to stuff it into the tube on the blunt end of a large twist-drill which just fits inside the tube. The foil is held in place by friction. I made a plug for the LED end of the tube by heating the tube with a gas torch and using the heated tube to melt a disc out of the end of a black plastic 35 mm film can. I made the holes for the LED wires with a heated panel pin.

The telescope equipped with finder is shown in Figure 3.2 (*overleaf*). The finder's components are assembled as shown in Figure 3.3 (*overleaf*). The precise positioning of the pinhole is achieved by clamping the finder and sighting at the night sky. The pinhole is properly positioned when there is no observable parallax between the pinhole and the stars when you move your head. The twist-drill makes a suitable tool for pushing the foil back and forth in the tube. The position of the LED is not critical, but obviously it should be close to the foil.

I made a mounting bracket out of aluminium strip. I drilled this to accept a toggle switch which is wired in series with the LED and the battery. I secured the bracket to the telescope with a Jubilee clip (pipe clamp), but all other parts are held together with Gaffer (Duck or duct) tape. It is a simple matter to align the finder with the telescope by bending the mounting bracket.

Figure 3.2 The telescope on an equatorial mount.

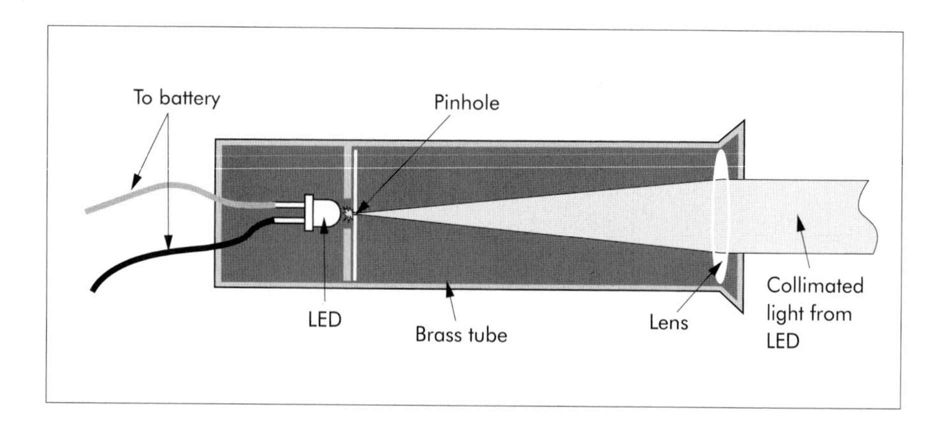

Figure 3.3 The LED finder.

The Completed Telescope

When I gave it first light, I was delighted by the pin-point star images I obtained at low powers. With a 32 mm Plössl (×16) it has a field over 3° in diameter and is a pleasure to use for exploring the night sky. My main area of observation is galactic clusters and this instrument is ideal for all but the smallest of these. Pushing the magnification beyond about ×60 begins to reveal the deficiencies of the objective lens, and the colour correction is not as good as that of a top-quality achromat, but I have seen a lot worse in some of the "low-end" refractors which are sold as astronomical telescopes.

My pleasure with this instrument is such that it has, apart from my binoculars, been my most-used astronomical instrument for the last 4 years. I initially mounted it on an altaz mount made from Meccano and old teleprinter parts. It then served a short time as an 11 × 80 finder until I sold the telescope it was on in order to finance another project, but since it has become my most-used telescope I have put it on an undriven equatorial.

An obvious extension to this project is to use the camera lens as a camera lens! The home-made focuser was inadequate to support the heavy old Canon FT which I use for astrophotography, so this has since been replaced by a commercial one, with adjustable tension, whose draw-tube doesn't fall out when the telescope is within 30° of the zenith.

Although it is heavy, its compact size makes it a very good "travelling" telescope and, since my back garden is lit by several streetlights and is partly surrounded by tall trees, I need an instrument I can easily put into the back of my small car. I am sometimes frustrated by the breakdown of the image at high powers, but I remind myself that it has already exceeded expectations and I do have other, less portable, telescopes which I can use for higher magnification.

Part II

Specialised Telescopes

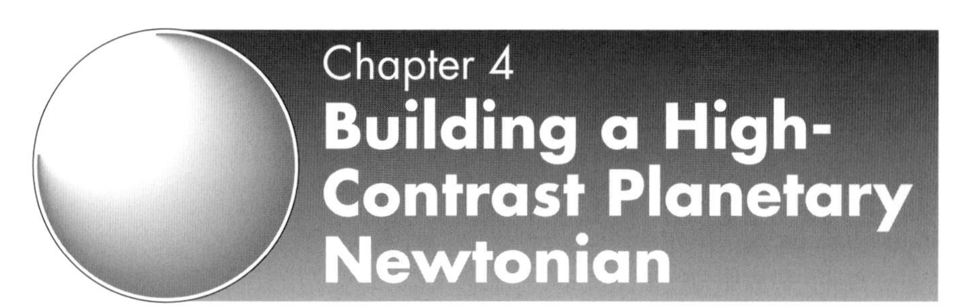

Chapter 4
Building a High-Contrast Planetary Newtonian

Gary Seronik

The reflector is much maligned as an instrument for lunar and planetary observing, but this 6 in (15 cm) f/9 Newtonian successfully uses every possible trick to increase contrast and improve image quality. This project typifies the way that ATMs adapt common techniques learned from earlier work in order to attain their aims in constructing a non-standard instrument.

Most telescope makers I know are on some kind of quest. They shape glass, trace rays, and formulate equations, all in search of an optical ideal that might never achieve solid form. For me, the "quest" is for the ultimate planetary instrument. As an observer, I find few sights as thrilling as Jupiter or Mars, resplendent with the kind of detail that can only be glimpsed when telescope and atmosphere are equal in perfection. While there is little a telescope maker can do to ensure atmospheric steadiness, there is something that can be done to ensure telescopic perfection. So began my personal quest – the quest that lead me to my present "ultimate" telescope: a 6 in f/9 Newtonian reflector (Figure 4.1, *overleaf*).

Designing Perfection

The detailed planetary views I sought would require a telescope with high contrast, good resolution and an optical train with a minimum of inherent optical aberrations. Since I planned on making my own optics for this telescope, I looked carefully at a number of designs

Figure 4.1 The telescope.

before concluding that the Newtonian reflector was not only the easiest to make but also the most likely to produce the views I was after. The closer I looked at other configurations, the more I came to appreciate the versatility of the simple Newtonian. What other design is capable of spanning so great a range of useful apertures and f-ratios? What other design offers such tremendous potential for obtaining optical perfection without a fully equipped optics lab? A Newtonian reflector it was to be.

Having settled on the Newtonian design, it was time to narrow the choice further and decide upon the aperture. It is my belief that satisfactory planetary resolution begins with a 6 in (15 cm) objective. Although a great deal can be seen with smaller scopes, years at the eyepiece have convinced me that a 6 in represents the threshold at which planets begin to reveal their most delicate details. It is a curious fact of telescopedom that while a 6 in refractor is thought of as a "serious" instrument for planetary study, a Newtonian of the same aperture is almost never regarded as more than a good starter scope. However, well-made examples of both are more alike than different, even when it comes to viewing the planets.

The next order of business was to choose the instrument's f-ratio. Next to aperture, f-ratio defines the performance characteristics of a Newtonian more than any other parameter. For the kind of observing I wanted to do, the choice was obvious – make the focal length as long as practically possible. Consider the benefits of the long-focus Newtonian: comparative ease of fabrication; a relatively large, aberration-free field of view; the opportunity to use a small secondary mirror; and a generous image scale which allows for the use of comfortable long focal length eyepieces. The greatest f-ratio I felt that I could handle easily was f/9. Much longer than this, and the instrument becomes unwieldy without significantly greater benefits; shorter, and the long focal length advantages begin to weaken.

Making the Primary Mirror

The heart of the Newtonian system is its primary mirror. Without a first-rate primary mirror, there is no way to produce a first-rate telescope – period. Having said that, there's no reason why anyone with determination cannot produce such a mirror the first time out. Perseverance and patience, rather than skill and aptitude, are the crucial ingredients for successful mirror making. The procedure for making a long-focus primary is basically the same as for any mirror, but there are a few significant differences to be aware of. Remember, the goal here is not simply a working mirror, but one suitable for use in a high-performance planetary telescope.

Although it is theoretically possible to produce as fine a mirror at f/4.5 as at f/9, in practice this seldom occurs for a number of reasons. First, it is far easier to make a long focal length mirror. The curve is much shallower and the amount of work required to bring such a mirror to the figuring stage is proportionately reduced. The second important advantage results from the nature of the Foucault test. This elegant yet simple analytical tool is a null test for spherical mirrors. What this means is that a perfectly spherical mirror will grey out evenly, or "null", during the test. Under these conditions, small defects in the mirror's surface stand out in stark relief. However, for all but the very longest focal length mirrors, the ideal shape for the primary mirror is not a sphere, but a parabola. Even though an f/9 parabola differs significantly from a sphere, we are still able to view minor flaws in the mirror's polished surface that would be invisible in a short-focus mirror. In this sense the Foucault test behaves like an "almost-null" test for an f/9 mirror. This is an important point when one considers that surface roughness is one of the main contrast killers in a Newtonian. If my experience with an f/9 parabola is typical, you should expect to spend as much time achieving a truly smooth surface as you will actually figuring.

Once a good sphere was achieved, I was ready to attempt the crucial step of parabolising. Since so little figuring needs to be done, the approach that worked best was the standard method of first producing a sphere, and then slowly working towards the parabola. If I didn't like the figure as it was emerging, I would simply go back a few steps to a sphere and begin again. Unlike shorter focal length mirrors that allow for correction on the fly, a long-focus mirror's figure appears so rapidly that there is almost no chance for alteration once it becomes evident. If one zone's correction lags behind the others, the chances are that over-correction will occur in another part of the mirror before the original zone is brought into line.

In my quest for optical perfection, I made close to a dozen parabolising attempts before achieving the final figure. Some of those attempts resulted in good mirrors that were just not quite good enough. I was after perfection, and although my patience and determination were tested, it was only a matter of time before I arrived at a figure that was as close to perfect as I was likely to produce. In the end, final testing showed the resulting mirror to be excellent – around 1/29th wave, although such figures can not be taken at face value. However, at the very least, I was confident that the mirror was far superior to anything I could reasonably expect to obtain from a commercial vendor.

Secondary Considerations

One of the main disadvantages of the Newtonian reflector arises from the fact that it is an obstructed system; the secondary mirror blocks a small percentage of the incoming light. This not only results in an image that is slightly dimmer, but, more seriously, one with reduced contrast. The amount of degradation depends upon the size of the obstruction – the larger the obstruction, the worse the effects. Beyond this simple truth, there is little agreement as to the extent of the harm. Some claim that obstructions smaller than 30% the diameter of the primary mirror inflict image damage that is barely noticeable, while others maintain that any obstruction at all is unacceptable. Most experts however, agree that if the secondary mirror's minor axis is less than 15% of the diameter of the primary mirror then the harmful effects all but vanish. Fortunately, one of the benefits of a long-focus mirror is that a relatively small secondary mirror can be used (see Figure 4.2). Since a secondary mirror of some size is necessary for the telescope to even work, the best one can do is keep this obstruction, and the resulting harm, to a minimum.

So how big should the secondary be? This too is the subject of great debate, and there really is no single "correct" answer. A good place to start is to figure out the size of a secondary that will just barely do the job. You can calculate the secondary's absolute minimum size easily. Simply measure the distance from the centre of the tube to where you want the focal plane to be – usually about $\frac{1}{2}$ in (1 cm) beyond the top of the fully retracted focuser. In the case of my 6 in this distance came out at about 5.25 in (135 mm). Take this number and divide it by the telescope's focal ratio. Dividing 5.25 (135) by the f-ratio of 9, I get a minimum

Figure 4.2 Comparative sizes of secondary mirrors in equivalent f/9 and f/4.5 systems.

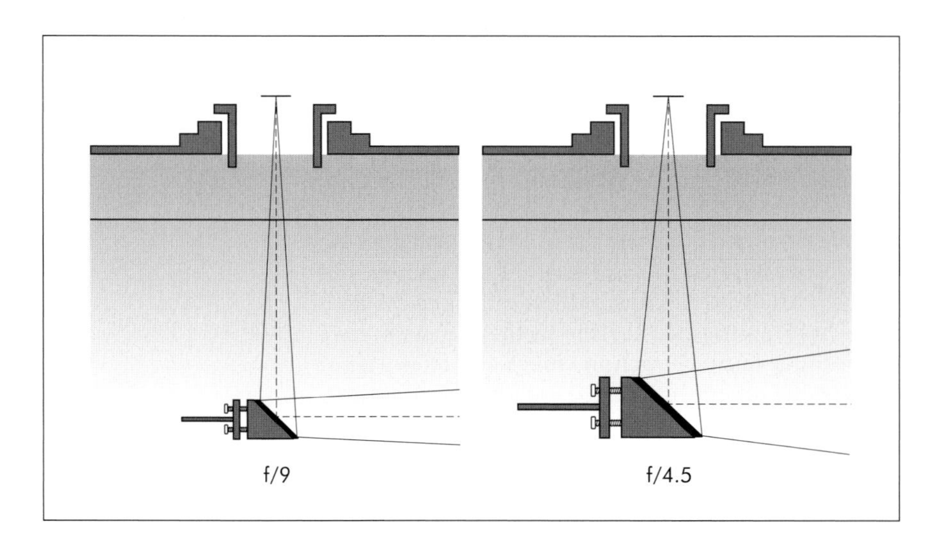

secondary size of 0.58 in (15 mm). A little experimentation quickly reveals the importance of a low-profile focuser. Indeed, it is virtually impossible to keep the secondary reasonably small if a standard rack-and-pinion focuser is used. Such a focuser often stands more than 4 inches tall and would require a secondary mirror nearly twice as large.

So, would a secondary mirror merely 0.58 in across really work? Yes, but there is a catch. With such a small secondary, only the on-axis rays will be deflected to the focal plane. What this means in practice is that any object that is not perfectly centred in the eyepiece will not receive the full benefit of the objective's light-gathering and resolution capabilities. A secondary of this size leaves absolutely no margin for error, so we would be wise to choose a diagonal that allowed for a larger area of full illumination. How much larger is the point at which opinion and personal preferences intervene in the design. However, for a high magnification planetary telescope there is little reason to have a very large fully illuminated field. My choice was made simply by finding the closest standard secondary size that was larger than the minimum necessary – in this case a 0.75 in (19 mm) minor-axis mirror. This provides approximately 0.18 in (4.6 mm) of fully illuminated field at the focal plane. Fans of low-power, wide-field eyepieces might desire a larger field of full illumination, but for my purposes I find this small secondary completely adequate. It presents a mere 12.5% obstruction, certainly not large enough to perceptibly alter the image.

Mounting the Mirrors

To eliminate all unnecessary sources of diffraction, attention should be given to how the primary and secondary mirrors are mounted. Traditionally, the primary is mounted in some kind of cell and held in place by three or four clips. If these clips project onto the mirror's aluminised surface, as they often do, they will produce faint diffraction spikes. A better way of securing the mirror, which doesn't involve clips, is to use flexible silicone rubber adhesive and glue the mirror directly into the cell. Three equally spaced pads, 70% of the distance from the mirror's centre, will hold the mirror securely once the silicone has cured. The procedure is simple: squeeze out three $\frac{1}{2}$ in (1 cm) blobs into the mirror cell, lay down $\frac{1}{8}$ in (3 mm) spacers (nails work well) between the blobs, set the mirror down gently (aluminised side up!), let the silicone cure, remove the spacers and you're done. If your mirror cell will allow for it, it's a good idea to support the mirror radially with three extra silicone pads, but for a 6 in this isn't strictly necessary.

To support the secondary mirror, a "spider" with three or four legs (vanes) is normally used. The amount of diffraction arising from such an arrangement depends on the thickness and number of vanes. Clearly, three vanes are better than four vanes – it's the total area of the obstruction that counts. However, the problem with straight vanes is that they tend to concentrate their diffraction into narrow streaks. This is what produces the familiar diffraction spikes around bright stars. One can make this diffraction less apparent by simply curving the vanes. This has the effect of spreading the diffraction out to such an extent that it becomes invisible. The stars lose their spikes and the image takes on a wonderful, refractor-like quality, minus the chromatic aberration.

For my 6 in I built a curved spider that consists of a single curving arc. The main design consideration is that the part of the spider that obstructs the mirror should be an even fraction of a 360° arc. For example, my single arc is really two 90° vanes joined end-to-end for a 180° obstruction – half of 360°. One could accomplish the same thing with three 60° arcs, or any other combination that adds up to either 180° or 360°. In practise, I found that absolutely precise arcs were not necessary. Almost any kind of curve worked to some extent, producing diffraction that was less noticeable than with straight-vaned spiders. I settled on the design pictured in Figure 4.3 (*overleaf*) because it was simple to make and produced about the same total diffraction as a three vane spider – except that the diffraction is rendered invisible.

Baffling

The best possible contrast occurs in a telescope with a high signal-to-noise ratio. Short of making the primary mirror larger, or the reflective coatings more efficient, there is very little one can do to increase the "signal."

Figure 4.3 The open end of the telescope, showing the curved spider.

The only option available is to beat down the "noise" level. In a telescope, the "noise" consists of stray light that somehow makes its way to the image plane, usually because of inadequate baffling. Given that proper baffling is quite easy to accomplish, and that the rewards can be simply astounding, it is surprising how little attention is paid to this in telescope construction.

In a nutshell, the point of baffling is to ensure that only light which has fallen directly on the primary mirror makes its way to the image plane. To accomplish this, stray light must be suppressed or, better, completely eliminated. Quite often, the only baffling done consists of a coat of matt-black paint applied to the inside of the tube. Although this is a good start, substantially better results are possible with only a slightly greater effort.

Some telescope makers take their lead from the refractor telescope and install a number of concentric ring baffles spaced along the inside of the tube. Leaving aside for the moment the question of whether these actually do anything worth while in a Newtonian system, such a strategy is bound to be more difficult to make and install than the alternatives. A second problem arises when such baffling is used along with a ventilation fan. Moving air is forced into the telescope's light path and exacerbates the effects of "tube seeing".

A much more simple and effective baffling approach is to roughen up the inside of the tube, thus creating millions of tiny baffles. This is the method used in my 6 in and I can personally attest to its effectiveness. It is a simple, three-step process:

1. Apply a generous coat of matt-black paint to the inside of the tube. (*Note*: It is worth testing different brands of such paint. I have seen a wide variation from one brand to another in both mattness and blackness, with some being almost semi-gloss.)
2. Before the paint begins to dry, throw several handfuls of sawdust or Slip-Not (a paint additive made of walnut shells crushed to the consistency of sand) down the tube. Distribute this material evenly by laying the tube on its side and rolling it over and over. Once the paint has dried, the majority of these particles will be firmly affixed to the tube wall. A few gentle taps to the tube should dislodge any loose pieces.
3. Apply a final coat of the matt-black. This is best done with a sponge and a gentle dabbing motion. Once this coat has dried your tube's interior will be wonderfully dark – so much so, that any missed areas will be glaringly obvious.

Once the tube interior is completed, it's time to look after the remaining baffling issues. How much remains to be done depends upon the specifics or your telescope, but there are three often-neglected problem areas. The first of these is the miscellaneous hardware inside the tube – the shiny nuts and bolts used to mount the focuser, mirror cells, or what have you. All of these can benefit from a coat of matt-black paint. It is also worth while to blacken the bevel on your primary mirror with a black chisel-point felt marker or a small brush loaded with matt-black paint. This is most easily done with your arm braced and the mirror slowly rotating on a Lazy Susan bearing or other turntable. Similar attention should be given to the secondary mirror's rim. Some secondary holders leave only the functioning aluminised surface exposed, but if yours does not, carefully go over the partly aluminised edge area that faces the focuser. If the very thought of going near these pristine shiny mirror surfaces with matt-black paint makes you break out in a cold sweat, skip these steps. There is no point in making a nervous wreck out of yourself for the sake of improved baffling!

Two remaining sources of stray light lie at either end of the tube. Look down the front of your telescope. Is it possible to see past the primary mirror and out of the back of the tube? If it is, you'll want to take care of this, since a substantial amount of light originating from this location can reach the focal plane. A ring of cardboard painted matt-black will usually take care of this. Keep in mind, that you don't want to simply cap the end of the tube, since some ventilation is necessary to allow the primary mirror to cool down to the ambient air temperature for optimum performance.

Finally, the most common and serious baffling problem results from stray light reaching the focal plane from over the top of the tube opposite the focuser. Telescopes equipped with low-profile focusers are especially vulnerable. Fortunately, this is a problem that has an easy fix: a longer tube. In the case of my telescope, this took the form of a removable tube extension or "dew shield" (Figure 4.4). You can determine how long this needs to be by putting your eye up to the racked-in focuser (without an eyepiece in place) and looking past the secondary, out of the top of the tube. With a ruler held against the tube opposite the focuser, you can see how long the extension will need to be.

Figure 4.4 The removable tube extension.

The Payoff

So does the view justify the effort? Absolutely. This telescope is the finest 6-inch I have ever looked through. With a high-quality Orthoscopic eyepiece, the views are remarkably contrasty and sharp. On nights of steady seeing the Moon presents an astonishing array of minute detail, bringing to mind astronaut Buzz Aldrin's description of a world of "magnificent desolation". Low-contrast details on Mars and Jupiter are revealed with the sort of clarity one seldom sees in instruments substantially larger. Deep-sky views are also rich with detail. Considering the benefits of high contrast on faint objects, this is not really surprising. After all, the very same characteristics which result in a fine planetary instrument also make for a first-rate deep sky telescope. This telescope has convinced me that an optically efficient 6 in telescope will outperform a larger instrument of lesser optical efficiency. It all results from attention to detail and determination – two key elements in the quest for optical perfection.

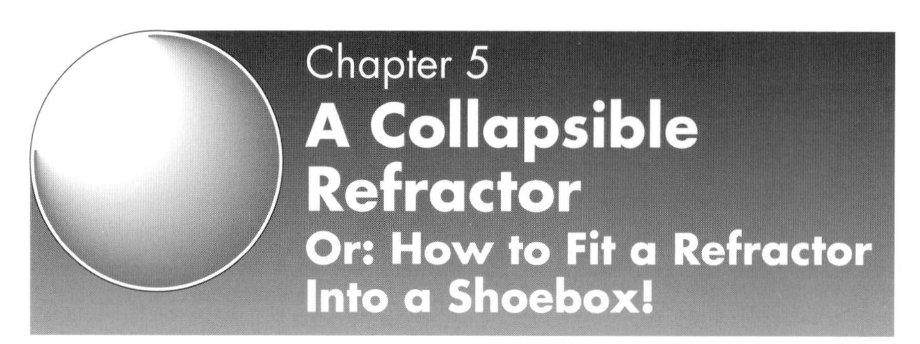

Chapter 5
A Collapsible Refractor
Or: How to Fit a Refractor Into a Shoebox!

Klaus-Peter Schröder

ATMs frequently evolve unusual designs in order to meet a specific need, in this case portability. Long focal ratio refractors are conventionally considered to give excellent images at the expense of being long and unwieldy, but this ingenious 110 mm (4.3 in) f/15 folded refractor of simple construction and outstanding optical quality can be packed into a shoebox for transportation.

The Concept

The conventional refractor is undoubtedly a fine instrument which produces high-contrast images. Among portable instruments, a refractor is therefore certainly the kind of telescope that offers most performance per inch of aperture – however, by no means per foot of tube length! But is there any way out of this trade-off?

Once I already had a nicely portable 15 cm (6 in) f/5 Newtonian, I started wanting an even more portable telescope, but one that was optically better. I considered the few options, but neither was I happy with a handy SC (Schmidt–Cassegrain) telescope, because of its compromised resolution and contrast caused by the central obstruction, nor did any fast focal ratio, apochromatic refractor appeal to me – especially when I looked at the prices! Complicating the problem even further, I also wanted to use that dream telescope with the same short tripod as my Newtonian. Therefore, any convenient focus would have to be at the upper end of the tube, like a Newtonian focus. And lastly, I deliberately wanted a long focal length because even a less sophisticated eyepiece performs really well over its full field when used with a slow f-ratio.

But, could such special requirements be fulfilled by any telescope at all? Happily, YES! And the solution, I came up with became my pet telescope: a long-focus refractor, which can fit into a shoebox!

What sounds so impossible is in effect not so difficult to make. The optics are simple: a long-focus achromatic lens which, simply by its concept, has negligible aberrations and a good colour correction, plus two optical flats. One is an elliptical flat of about 35 to 50 mm ($1\frac{1}{2}$ to 2 in) minor-axis diameter, as mass-produced for Newtonian telescopes. The other one must be – it's the only non-trivial optical component – a very good flat with a diameter of about two-thirds of the lens's aperture. These flat mirrors simply "fold" the focal length in the way sketched in Figure 5.1, without introducing any additional aberrations or critical dependence on collimation, but cutting down the tube length to about 40% of the focal length. The eyepiece is in a Newtonian-type position, but there is no central obstruction.

This concept of a "folded" refractor was quite common among the few German ATMs in the sixties, since it combines most of the advantages listed above. Also, in a Dobson mount the performance of this type of telescope is just great – I saw such a telescope about 20 years ago in the home of a retired physicist and late Hamburg ATM. He constructed it long before that kind of mount became known as a Dobson.

However, such a "folded" refractor still does not fit into a shoebox. Nevertheless, I got hooked by the simplicity of this design. Also, the combined costs of the optics are much less than the price of a same-size, fast f-ratio apochromatic lens and are comparable to a same-size (but obstructed) SC telescope. Not much later, in an optical company stock-clearance, I got a reasonably priced lens of 11 cm (4.4 in) aperture and f/15, a 50 mm (2 in) elliptical and a 75 mm (3 in) round flat, all of good quality. There I was!

When I thought of the tube construction, whether it should be a long box (of plywood, maybe) or rather based on a plastic tube, it occurred to me that except for the 75 mm flat all the optical components, namely the lens, the elliptical mirror, the eyepiece and the

Figure 5.1 Sketch of the optical light-path in a "folded" refractor with Newtonian-type focus position, as seen from the side (not drawn to scale) and from the front.

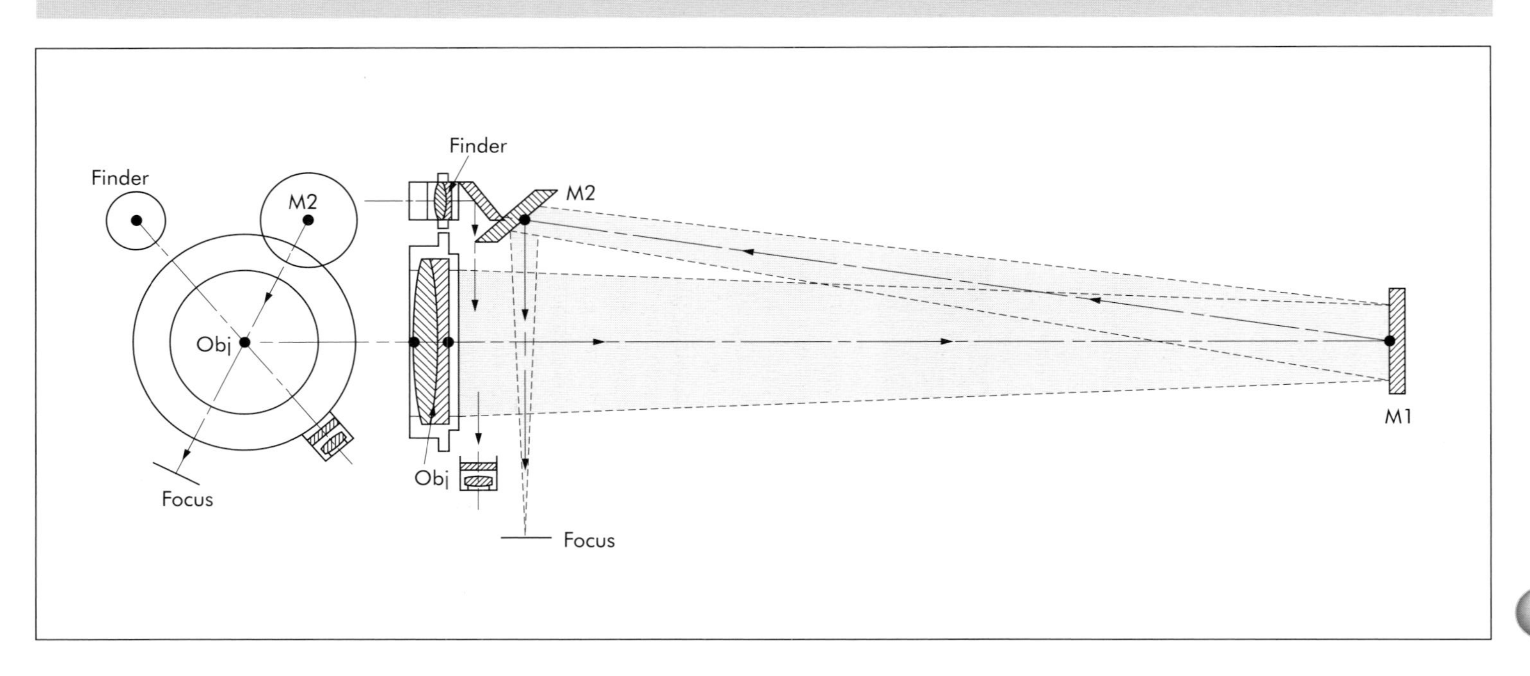

finder scope, could be placed within only the volume of a shoebox! That made my decision in favour of a box. But what to do to bridge the distance to the round flat? At that stage I had two ideas: (1) to bridge that distance by four small tubes or poles of 0.45 m (18 in) length which, for structural stability, would converge from the rear box-corners towards the smaller diameter of the mirror cell, and (2) to integrate the finder scope optics in the box, using the optical arrangement sketched in Figure 5.1. The concept of my collapsible refractor was born: all you need is a box, four poles and a 75 mm (3 in) mirror in its cell and you have a high-quality refractor of 1.65 m (65 in) focal length and with an integrated finder scope!

Construction and Materials

Since, in most cases, optics are not bought from the shelf but obtained as and when they become available, the following description is intended to provide the interested reader and ATM with a general guide to making the best practical use of his or her particular optics, tools and skills, rather than a strict recipe. The way things worked out for me (see Figure 5.2) might not be available to everyone else, but there are really many other good ways to reach the same end.

I did most of my constructing with aluminium profiles which I obtained in a DIY store, and which I was able to cut precisely with a mitre saw. I screwed the connections, gluing them with slow-curing epoxy, which gave me time to check all right angles and so on of the whole construction. For such materials, you also need some appropriate files and a mounted drill. However, you can facilitate your work a lot if you settle for a box made of 4–5 mm ($\frac{3}{16}$ in) good-quality (marine) plywood, with reinforced corners, and if you have a friendly DIY store that provides cutting to customers' specifications.

To provide decent cut-outs for the front side lens-fit and for the rear side light baffle (you can work out its geometry with another 1:1 sketch and use a thin sheet of good plywood), you also need a fretsaw or coping saw. Otherwise, only ordinary tools are required – a large pair of pliers, screwdrivers and so forth. More

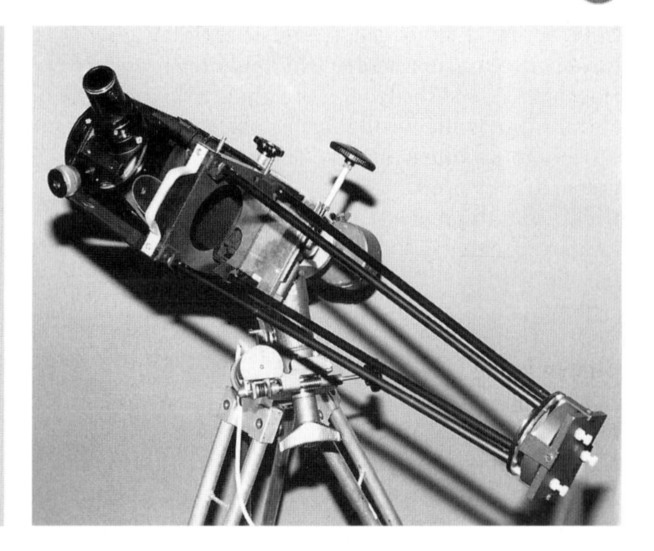

Figure 5.2 The collapsible 11 cm f/15 refractor constructed by the author, shown without the black fabric cover which keeps out stray light. The front box is an anodised aluminium construction, but plywood painted matt-black would do equally well. The complete tube assembly weighs 7 lb (3.2 kg).

important, really, is a good helping of common sense, enough patience and the capacity to enjoy DIY; with these, you will certainly get there.

To start with, make full-scale (1:1) drawings of the light path, like that in Figure 5.1, and use it to decide the best positions of your flats and finder scope optics, and the size of your front box. It should be long enough to go about 5 cm (2 in) beyond the centre of gravity. Otherwise, it will be difficult, with any mount, to have the telescope well balanced and rigidly attached. The stiffest part of the telescope is its front box, and it should coincide with the declination (or height) axis of the mount. You should estimate the weight ratio of the round flat, including its holder and the tubes, over the sum of all the other components; you will find that this is a small figure. Multiply it by the distance between the lens and the round flat, add about 5 cm, and you have the approximate centre of gravity, measured from the front lens. Plan the rear of the front box to go at least 5 cm beyond that point and to be reinforced by a frame.

For focusing, I bought a Newtonian low-profile helical 2 in focuser. Together with the 50 mm (2 in) elliptical flat, it permits use of really wide-field eyepieces – such as a 32 mm Erfle with 35 mm ($1\frac{1}{2}$ in) field of view, which gives me a full degree of very crisp stellar images in the sky! My space-saving design sends the focus into one corner, and a Newtonian focuser could well ride on an appropriate cut-out from a wooden box. In my telescope, as seen in Figure 5.2, I have screwed it to the front side, sideways. This, however, requires an edge which is very precisely cut (machined), and the front to be made of a stiff sheet of aluminium, which probably goes beyond the easy-to-do things for most ATMs. The finder scope can be focused and aimed with its diagonal mirror – a bit tricky, but if that small diagonal is well supported and mounted, then you'll do this job only once. If you don't like to bother with such a mount, you can buy two commercial Newtonian diagonal holders, one for the finder scope diagonal and one for the elliptical flat of the telescope.

The round flat at the lower end of the telescope needs an adjustable mount which attaches to a front-ring, as in my construction, or it can be housed in a small, flat box which has an appropriate opening in front. The mirror mount would work much like one for a Newtonian main mirror, but those are only available from 10 cm (4 in) at best, and it is not so difficult to make one yourself: an either square or round piece of plywood (the same size as the flat) can form the base of the mirror mount and the flat is held in place by four retaining clips made from a flat aluminium strip. I pre-positioned the flat into approximately the correct angle by shimming it with padded sticky fixers on the front-side – which also keeps the mirror free of mechanical stress. Such sticky fixers provide sufficient flexibility for fine adjustment by three plastic screws, which meet the 75% zone from the back of the flat.

Finally, you need four poles or tubes to connect the front box with the rear flat. You get the proper length from your 1:1 construction plan – it will be about 30% of the focal length. If you provide sufficient inclination (at least 10°), with which these poles or tubes converge from the front box rear corners down to the front of the mirror housing, you will find that 10 mm ($\frac{3}{8}$ in) threaded rods or 12 mm × 1 mm thick ($\frac{1}{2}$ in × $\frac{1}{12}$ in) aluminium tubes already provide enough stiffness. The aluminium tubes are much lighter and their ends can be reinforced by M6 screws for operation with an Allen key. These usually have a round head with a fitting diameter of 10 mm, and the threaded ends can stick out as much as required. If you manage to cut the tubes to exactly the same length, collimation would not change when they get screwed on in a different order. To provide the right inclination on the facing sides of the front box and the mirror housing, I cut pieces of a U-profile (aluminium or brass), bent them slightly with a strong pair of pliers, drilled the required holes and mounted them with their openings showing outwards. That gives an easy accessibility of the nuts, which fix the tube-ends. If you go for a machined front-ring on the mirror-housing, then you can produce the right inclination on a lathe, as I did with my construction (see Figure 5.2).

After completing the construction you may worry about collimation. But that is as uncritical as with a long-focus Newtonian telescope. Centre a 6 mm hole in your focuser and work your way down the light path: Do you see the elliptical diagonal in the right position? Do you see the round flat centred in it? (If not, tilt the elliptical.) Do you see the lens centred in the two flats? (If not, tilt the round flat.) Finally, make sure the lens sits square to the optical axis. You can test that, as with any conventional refractor, by observing the reflections of a small bulb (torch) which you hold on-axis. To keep your eye on-axis as well, you can use a small, cheap mirror: scratch off a central spot of its coating, and fix it to a short, diagonally cut 1.25 in tube (or 2 in, whichever is your focuser's ID) with a side-opening, centred over the mirror. The torchlight can now enter on-axis from the side, while your eye is centred at the same time. If the lens is tilted, its reflections of the bulb (four for a doublet) will not be in the same place but will form a line pointing to the side on which the lens is tilted towards you.

Finally, you need to cover the whole tube-assembly with a fitting black sheet of fabric to keep out any stray light, put it on a mount – and enjoy!

This description may have inspired you to endeavour your own construction of such a handy and portable refractor. I, for my part, was very pleased with the result. I was positively surprised at the absence of stray light, and there is no reduction of image quality from the two extra optical surfaces. The contrast is as excellent as with a conventional refractor and definitely better than any of my non-refractors, including even my Questar. I observed the Moon at a magnification of 240× and the image was still crisp; the centre of the Orion nebula appeared best at 160×. My collapsible refractor easily outperforms my much heavier 15 cm f/5 Newtonian telescope in terms of resolution and contrast, and despite its smaller aperture it does not lose on deep sky objects either! The images are a bit darker, but of a better contrast and are more pleasant to look at.

Now, honestly, does my collapsible refractor fit into a shoebox? Well, almost. It needs a box for hiking boots (but that takes the rear mirror as well), and the four 12 mm tubes go extra. That's good enough for me!

Chapter 6 The Construction of a Buchroeder Quad-Schiefspiegler

Terry Platt

Some instruments are just not available off the shelf, but somebody with ATM skills can use these to avoid the high cost of having such a specialist instrument custom-made by a manufacturer. Terry Platt's telescope is an excellent example of an instrument being built to meet a specific need which could not be met by commercially available instruments. He uses this 318 mm (12.5 in) f/20 unobstructed reflector for planetary CCD imaging.

Origins

Although the advent of high-quality commercial telescopes has made home construction a minority interest, there are some optical systems which cannot be bought at the local dealers! The various kinds of off-axis reflector are in this category, and virtually all of those now in use have been built by enthusiasts, working in their spare time with basic materials. I have been grinding mirrors for telescopes since about 1963, mostly for Newtonian reflectors of up to 460 mm (18 in) aperture, but when a design for a large, unobstructed reflector was published in a magazine, it immediately caught my attention. The magazine was the (unfortunately) now defunct *Telescope Making*, edited by Richard Berry and published during the 1980s by Kalmbach in the USA. The specific issue was Fall 1986, no. 28.

At that time, I was beginning to record high-definition planetary images with a prototype CCD camera, which was destined to become the Starlight

Xpress frame store system, and I was finding that it was limited by the telescope performance. This was a 300 mm (12 in) f/5 Newtonian, which was not especially well suited to planetary imaging, mainly owing to the fairly large flat mirror and spider assembly. The attraction of owning an unobstructed long-focus reflector was difficult to resist, and so I resolved to have a go at building one, based on the design in the magazine.

Off-Axis Telescope Design

The first "off-axis" reflector design was the Herschelian (Figure 6.1), employed by William Herschel for many

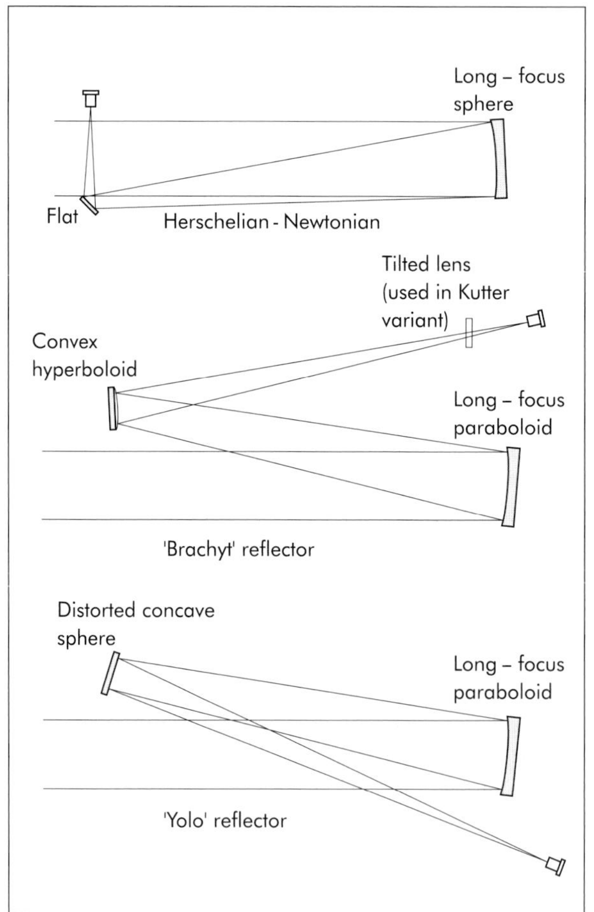

Figure 6.1 Some unobstructed reflectors.

ground-breaking observations with his larger telescopes about 200 years ago. The Herschelian is a simple, one-reflection instrument, in which the observer looks down at the primary mirror, using an eyepiece mounted at the top edge of the tube. Herschel employed it mainly to overcome the light loss associated with a Newtonian flat (the metal mirrors being of rather poor reflectivity). However, such simple systems have inherent astigmatism and coma, which can be reduced only by using an excessively long focal length primary mirror – not good for a compact observatory!

Since Herschel's time, many attempts have been made to design a simple and compact off-axis reflecting telescope. Optical losses are no longer an objection to building a complex off-axis reflector, and the lack of an obstructing flat and spider assembly is a strong attraction, as this improves planetary contrast and definition by removing these causes of diffraction.

The problem with all such systems is in the adequate correction of the extreme aberrations caused by tilting the optics. This generally leads to designs which involve astigmatic secondary mirrors, and/or extremely long tube assemblies, both of which are inconvenient or difficult to make. The first practical systems were developed in the mid-1800s and often known as "Brachyt" reflectors (from the German for "broken"). These are two mirror devices with the aberration of the tilted long-focus primary being corrected by the opposite aberration of a tilted convex secondary. This design works well, but is practical only in small apertures of up to about 125 mm, being limited by coma and residual astigmatism.

The Austrian optician, Anton Kutter, was a leading proponent of off-axis instruments and produced many improved designs with large apertures during the 1940s and 50s. These were mostly two-mirror designs in which a weak, tilted, cylindrical lens was used to correct the residual coma and to flatten the image plane. Kutter coined the term "Schiefspiegler" (from the German for "tilted mirror") for these telescopes and the term has become widely used to refer to any tilted optics reflector.

Kutter's schiefspieglers are superb planetary telescopes, but are still rather long and cumbersome.

It is possible to distort one mirror to correct residual aberration and the early 1980s saw a new design, the 'Yolo', which uses this method. The Yolo is a design by Arthur Leonard of Yolo county, California, and uses a stressed spherical concave secondary in an off-axis Gregorian configuration. Unfortunately, Leonard's Yolo is still a long instrument and the stressed secondary is not an appealing feature! It was not until the mid 1980s that off-axis designs became reasonably compact.

The breakthrough that makes them really practical for the small observatory, is the replacement of the correcting lens with a third tilted mirror, which performs the coma correcting function and folds up the optical train to make the instrument a more manageable size. Dick Buchroeder of California, amongst others, refined this design and generated the parameters which enable a relatively compact Tri-Schiefspiegler to be built with an aperture as large as 320 mm.

The issue of the magazine already mentioned was where I first met this telescope format. There two versions of the Buchroeder system were featured, in a 12.5 in (320 mm) aperture. One of these was a classic Tri-Schiefspiegler by Steven Johnson of Baton Rouge, Louisiana, and the second was a slightly modified design by Richard Wessling of Milford, Ohio. This second design was particularly interesting, as it added a small flat mirror to project the final focus into a very convenient location at the upper telescope tail, and this also resulted in a "normal" image orientation (the three-mirror design having a laterally inverted image). I decided that this four-mirror design was almost ideally suited to my need for a high-performance planetary reflector – and the plans began to unfold in my mind!

Figure 6.2 The Schiefspiegler optical layout.

The Wessling/ Buchroeder Quad-Schiefspiegler

A diagram of the optical system of the QSS is shown in Figure 6.2. The primary mirror is a 318 mm (12.5 in) diameter ellipsoid (55% of parabolic), with a radius of 7620 mm (300 in), and is tilted at 3.15° with respect to the incoming light The converging beam is intercepted by a 150 mm (6 in) diameter convex spherical secondary, 2172 mm (85.5 in) from the primary and tilted at 9.6° to the axis of the cone from the primary. The newly

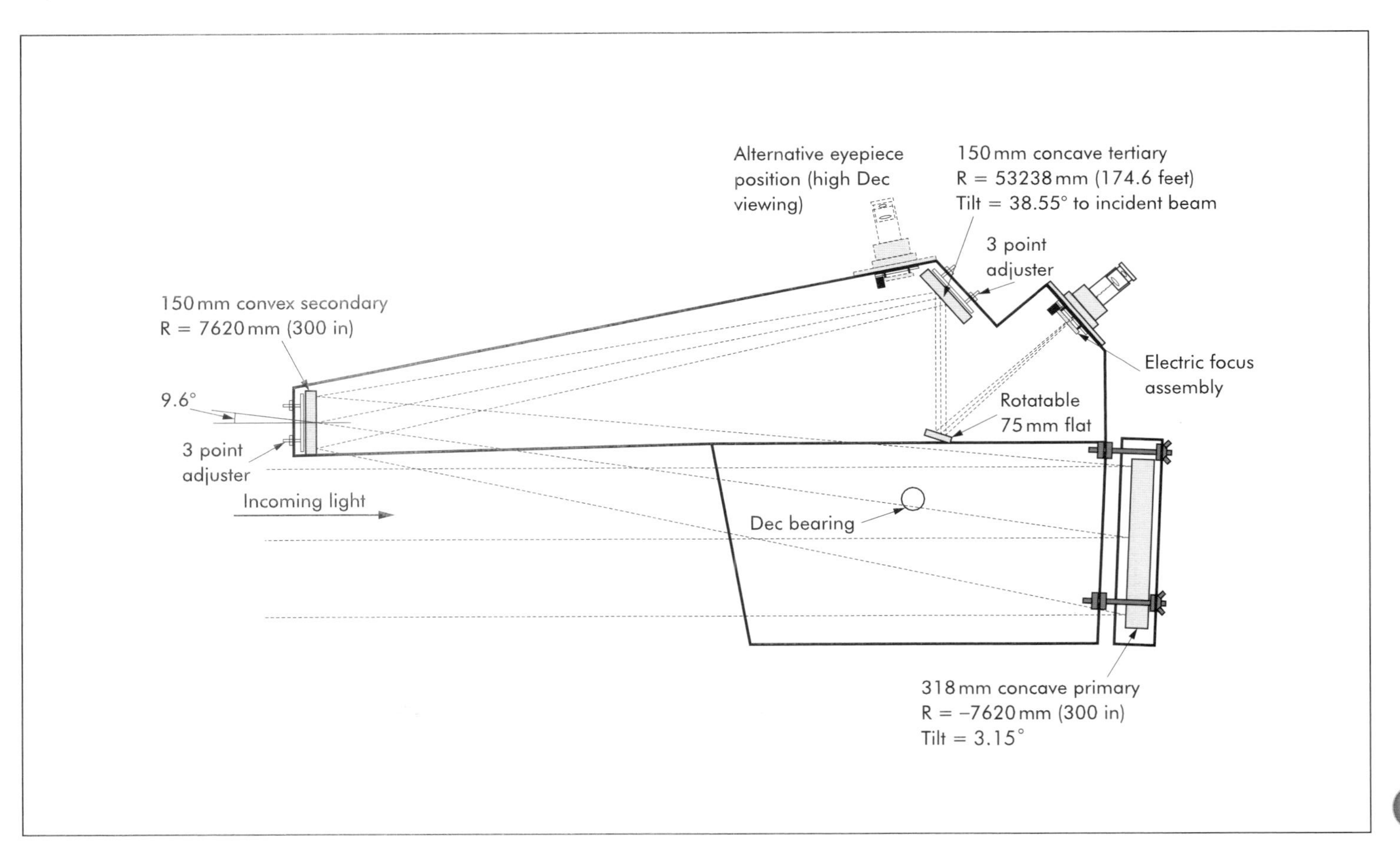
Alternative eyepiece position (high Dec viewing)
150 mm concave tertiary
R = 53238 mm (174.6 feet)
Tilt = 38.55° to incident beam
3 point adjuster
150 mm convex secondary
R = 7620 mm (300 in)
Electric focus assembly
9.6°
Rotatable 75 mm flat
3 point adjuster
Incoming light
Dec bearing
318 mm concave primary
R = −7620 mm (300 in)
Tilt = 3.15°

extended convergent beam then travels back down the tube assembly and is intercepted by a weakly concave tertiary mirror, very strongly tilted at 38.55° to the beam. The curvature of the tertiary is also spherical, but with the extremely long radius of 53 238 mm (174 ft 8 in)! The output beam is now directed across the incoming optical axis of the instrument and would come to a focus on the opposite side of the tube in the conventional design. However, in the quad design, a small, flat mirror is placed in this beam, just above the incoming light path, and directs the light to come to a focus at the upper tail end of the instrument, where it can be very conveniently observed with an eyepiece, or camera. The focal plane is actually anamorphic, that is, tilted with respect to the optical axis, in this case by 9°, but this does not have any obvious effect on the image quality across the eyepiece field.

The main difficulties encountered when building this telescope were (1) testing the convex secondary, (2) making and testing the very weak tertiary, and (3) holding the entire assembly in good collimation! Here is how these problems were handled:

The Primary Mirror

The smoothness of the primary mirror of the QSS is crucial to its performance, and so I decided early on that this should be made from Pyrex, rather than soda-lime glass. This gives it a better thermal stability during polishing and reduces any tendency for "dog biscuit" irregularities. At the time (1988) there were no convenient sources of Pyrex blanks in the UK and so I ordered a 318 mm × 50 mm (12.5 in × 2 in) blank from Willmann-Bell in the USA. The price was quoted as $150, which seemed very reasonable, but I had not accounted for the dreaded British Customs, who promptly slapped on an extra £48 as import duty! (Please take this as a warning about importing items to the UK from the USA.)

I am always keen to save money, if it does not compromise the performance of my telescopes. (I was especially keen to do so after my experiences with the Customs people!) I therefore decided to make all of the smaller optics from ordinary 18 mm ($\frac{3}{4}$ in) plate glass.

The convex secondary mirror of this instrument has exactly the same (inverted) radius as the primary, and so it is convenient to make the secondary from the centre of the primary's tool disc. It seemed unwise to cut the tool after grinding the primary, as stresses would probably be released and, also, an accident could easily occur during cutting, so I built the tool up out of individual glass segments. First, I bought a 318 mm × 12 mm thick (12.5 in × $\frac{1}{2}$ in) circle of float glass from the local glass merchants for use as a backing sheet. I then cut a 150 mm (6 in) diameter secondary mirror blank from 18 mm float glass and cemented onto the centre of this backing disc with soft pitch. Eight roughly cut sectors of 18 mm plate were then arranged around the secondary blank to complete the tool disc and also cemented down with pitch. Next, I levelled the whole assembly of blocks by gentle heating while in they were in contact with a weighted glass plate. This caused the pitch to soften and allow the blocks to become co-planar after an hour or so.

I undertook the grinding of the primary mirror in the usual fashion, using the fabricated tool mounted on a slowly revolving home-made motorised table. The primary mirror radius required by this telescope is 7620 mm, which is long (f/12) and results in a relatively short period of coarse grinding to reach the desired depth. I used a simple dial-gauge-based, home-made spherometer to monitor the radius and then refined the measurements by optical testing of the wet blank as the correct value was approached. As the design is sensitive to errors in the radii of the mirrors, I spent some time in adjusting the primary by grinding with 220 grit until the optical tests gave a radius of about 7700 mm (303 in), the remaining 80 mm (3 in) being the allowance for fine grinding. The coarse grinding over, I gently bevelled each of the tool blocks with a sharpening stone, so that the chances of glass chips being released from the sharp edges (and causing scratches) was much reduced.

I carried out the fine grinding of the primary using a sequence of 220, 400 and 600 grit carborundum, followed by washed and deflocculated 700 emery. The emery was kept in suspension in a squeezy bottle with a small amount of detergent to promote the breakdown of small clumps (floccules) and the elimination of larger grains by precipitation. About 2 hours with each carborundum grade sufficed to remove the pits from the previous smoothing sessions, but I spent 4 hours with the emery to give the finest possible finish.

The shallow curve of the primary mirror made it possible to mould a pitch lap for polishing onto a spare flat glass blank. I needed the original tool disc for making the secondary and so I didn't want to risk damaging its surface and, if any sleeks needed grinding off the primary, I had easy access to the tool, without destroying the polisher. I started the polishing with a full-size tool, but the mirror was found to be slightly hyperbolic and so I made up a 100 mm (4 in) polisher on a varnished wood disc and used it to reduce the 70% zone until the mirror was nearly spherical. After this, the full size tool was all that was needed to achieve a high polish and an accurately spherical surface. The long radius makes the Ronchi and Foucault tests very sensitive to surface errors and you can easily see where corrections are needed. Several weeks after starting work, I now had the test sphere for the secondary!

The Secondary Mirror

Like most amateurs, I have made very few convex mirrors and do not possess any large test spheres suitable for an autocollimator test rig. Fortunately, the secondary of the QSS is deliberately designed to be tested against the primary by interference fringes, and so all that is needed is a monochromatic light source and three pieces of foil to separate the surfaces! Although the test is simple, the monochromatic light source can be something of a problem, as bright, wide-angle sources tend to be expensive laboratory equipment. It is certainly possible to get results from a neon night-light bulb, but you need to be in a well darkened room to see the fringes clearly. Sodium lamps are almost ideal as light sources for interference testing and, with this in mind, I asked the local street light maintenance department if they had any used sodium tubes and ballast chokes. I was surprised and pleased to be given 4 old tubes and a ballast from a street lamp which had been demolished in an argument with a lorry!

I installed the choke and a 48 W tube on the ceiling above the bench in my workshop and, after the 10-minute warm-up time, a bright golden glow was flooding the entire room – no darkness is needed when interference testing with a sodium lamp! I rested the primary mirror on a carpet tile for distributed support and then placed three postage-stamp-sized squares of aluminium foil on its surface in an equilateral triangle of the same general size as the secondary mirror. These provide a small separation of the glass surfaces during testing and prevent scratches, which can be caused by trapped particles. When the secondary was gently placed on the aluminium spacers and gently aligned by pressing on the back surface, a beautiful bright set of interference fringes came into view.

In theory, the primary and secondary curves do not have to match very closely for the telescope to work, but interference testing demands a match to a fraction of a wavelength, so that irregularities can be detected in the fringe pattern. Inevitably, the polishing of the primary will push it several waves away from the secondary's curve and so I spent the first few hours with a deformed lap, polishing the secondary to a matching radius. During this time, I performed regular checks of the curve under the sodium light and dealt with developing bumps and holes before they became severe. Although the secondary needed several waves of correction, I found it surprisingly easy to get it to match the primary to better than a quarter wave. This is due largely to the clarity of the interference test, which gives you a precise indication of the errors and their magnitude and so simplifies the work considerably. If there is any doubt about the direction of an error, you just place your thumb on the secondary surface for a few seconds and replace it on the test bed. The thermal bump caused by your thumb then indicates the direction of any other defects!

About 10 hours of corrective polishing eventually resulted in a secondary curve which I felt was spherical to within better than a tenth of a wavelength of sodium light.

The Tertiary Mirror

This telescope has one of the weakest optical surfaces that I have ever come across in any design! The tertiary mirror is used only to correct for residual aberrations and it needs only a very weak curvature to do this, resulting in a 150 mm concave sphere with the huge radius of 53 238 mm. The actual active diameter is only about 80 mm (3.15 in), but it is difficult to preserve a good edge on very weak optics and so it is easier to make the mirror oversize.

Testing such a weak mirror is a very difficult problem, as the light source for a Foucault test needs to be very bright, when so far from the mirror, and the test shadows are almost invisibly small when seen from 53 m (175 ft) away! Fortunately, such a long focal length makes the Foucault test extremely sensitive and some of this can be sacrificed to create a very bright light source. I decided to try to use the very thin filament of a 12 V, 5 W "festoon" lamp as an intense simulated slit light source, and this proved to be very successful. I made a test assembly from marine plywood, which consisted of a 300 mm square base plate carrying a deep, shielded box at one side. The festoon lamp was mounted within it and a knife-edge device, made from an aluminium plate screwed onto a small movable wooden base, was positioned adjacent to it. The test set was completed by mounting one half of a spare pair of old binoculars at the back edge of the base plate. This monocular made it possible to see the mirror surface clearly from the 53 m test distance!

I had no way to test this mirror indoors, and so all the checking had to be carried out during evening twilight in an adjacent grassy field. I mounted the mirror on a bracket fitted to the top of a photographic tripod, and screwed the test assembly onto a similar tripod at the specified test distance.

The curvature of the tertiary is so slight that I had to carry out the coarse grinding with 220 grit carborundum and measure the minute 0.053 mm (0.002 in) sagitta of the mirror approximately with very thin feeler gauge fingers! Needless to say, the radius of this mirror was very sensitive to temperature differentials, and a change of half a metre (20 in) could happen between first setting up to test and a second test made after 10 minutes of cooling. I worked through many brief periods of normal and inverted fine grinding before the radius was to my satisfaction, but at least the surface finish was excellent for rapid polishing!

Although the test setup was very unconventional, I found the mirror to be fairly easy to figure to a spherical surface. The crucial factor was to work very gently and slowly, with very little pressure and well-washed cerium oxide rouge. Aligning the two tripods was always a long-drawn-out performance, and I often set them up just as it began to get dark and left them in place until late at night. At intervals of about 20 minutes, I would take the mirror out, place it on its stand, check the figure and then disappear into the workshop again. This worked well until the night when I want out to test the mirror – and found that someone had stolen the tripod! After that, I kept a very wary eye on the test set and would check every few minutes, just to be certain that another tripod didn't vanish the same way! Fortunately, we had a long spell of fair weather at the time, and so the figuring progressed quite quickly. I managed to generate an excellent spherical surface within about 2 weeks of starting work and confirmed this by critical study of the returned image with a short-focus eyepiece. It is interesting to note that the polishing of this mirror shortened its radius by about 300 mm (12 in) – a sensitive surface indeed!

The Mechanical Design

The Quad-Schiefspiegler is an unusual telescope, and requires an unusual tube assembly. As the optics are not coaxial, the system has a considerable vertical depth and looks decidedly un-telescope-like! I am not equipped to handle large sections of metal and prefer to work in wood, which is also generally lighter and less expensive. The main disadvantage of a wooden telescope is in the dimensional sensitivity to atmospheric humidity, but a good coat of paint or varnish offsets most of this.

I first drew the optical system to scale on a large sheet of graph paper and then sketched an approximate outline around it, allowing for declination bearings, and a switchable eyepiece location for use at high and low declinations. It quickly became evident that the main tube assembly did not need to be very long, and that instead a long, narrow nose would be adequate to carry the secondary mirror. The maximum vertical cross-section is coincident with the position of the tertiary mirror and is close to the centre of gravity of the telescope as a whole. Years ago I was quite active in building and flying radio-controlled model aircraft and so I tend to think in terms of aircraft fuselages when designing a telescope tube. My design, therefore, gravitated towards a plywood skinned series of bulkheads, arranged in strategic positions. The final result can be seen in Figure 6.3 (*overleaf*), which shows that the tube is arranged as two vertically stacked rectangular boxes. The lower box is approximately 380 mm (15 in) square in cross-section and is composed of a 9 mm ($\frac{3}{8}$ in) plywood skin applied to a series of three12 mm ($\frac{1}{2}$ in) plywood formers. This carries the

Figure 6.3 The "back end" of the telescope, looking up out of the observatory dome.

primary mirror cell at the lower end, which is attached by three lengths of 10 mm ($\frac{3}{8}$ in) studding and locknuts for mirror alignment, and is tilted at 3.15° to the longitudinal axis. The three large formers have smaller rectangular extensions on their upper edges and these become the internal formers for the lower end of the upper box section. This upper box is the enclosure for the secondary and tertiary mirrors of the Schiefspiegler, and also carries the eyepiece assembly. It runs the full length of the optical system and is composed of a 6 mm ($\frac{1}{4}$ in) plywood skin over 9 mm ($\frac{3}{8}$ in) plywood formers, except where the 12 mm ($\frac{1}{2}$ in) plywood extensions project into it. The secondary-mirror-carrying nose is an extension of the upper box, and is reinforced at intervals with 9 mm plywood rectangles with central apertures for the optical path from the secondary mirror to the tertiary. Because of the shallow angle made by the converging beam from the primary mirror, a long, truncated elliptical aperture in the lower surface of the upper box is needed for some distance ahead of the secondary, the rest of the underside being sealed by a sheet of 6 mm plywood. I finished the interior of the tube assembly in matt-black paint and externally in a two-tone combination of gloss white and deep blue oil-based paint.

This entire structure is held together by wire nails and PVA adhesive and is exceptionally strong and light. It is, however, rather long, and so the locus of the secondary mirror support is kept within a minimum volume by the use of a fork mounting. This allows the entire instrument to be housed within a 3.95 m (13 ft) dome (Figure 6.4), the maximum size that can be accommodated on top of my workshop! I made the fork itself from exterior plywood, in this case 20 mm ($\frac{3}{4}$ in) thick, and heavily reinforced it by the use of external gussets and a box section at the lower end. The tines are capped by short lengths of Dexion slotted angle-iron, which provide strong and adjustable mounts for the pillow block ballrace declination bearings. These are standard Picador types, with a 20 mm bore, and carry short lengths of 20 mm steel studding (all-thread) with locknuts, which act as the declination shafts.

I provided the Schiefspiegler with a tangent arm declination drive system, which consists of a 760 mm

Figure 6.4 The author's telescopes. The Schiefspiegler sits above a 330 mm (13 in) f/4 deep sky camera.

(30 in) long, 6 mm ($\frac{1}{4}$ in) thick sheet aluminium arm in pressure contact with a 250 mm (10 in) aluminium disc, which is screwed to the telescope side wall. The disc is packed out from the wall by a 20 mm ($\frac{3}{4}$ in) plywood circle, and aluminium clamps with synthetic leather liners grip its edge. I can set the friction so that the telescope can be slewed in declination by gentle pressure, but remains locked in position against the tangent arm when left alone. The lower end of the tangent arm is driven by a stepper motor and lead screw assembly, which allows for approximately 20° of declination adjustment.

Below the fork is the RA drive and bearing assembly (Figure 6.5). I made most of this structure from 75 mm × 75 mm × 6 mm (3 in × 3 in × $\frac{1}{4}$ in) mild steel angle, welded onto a 500 × 750 × 12 mm (20 in × 30 in × $\frac{1}{2}$ in) steel base plate. I bought all of these parts at a local scrapyard for a few pounds and assembled them at home with a portable welding set. The Polar shaft is a 100 mm (4 in) diameter steel tube with 10 mm ($\frac{3}{8}$ in) walls and is welded to a 400 mm × 12 mm (16 in × $\frac{1}{2}$ in) steel disc at the upper end. I turned this disc true on a lathe and then bolted it to the fork by four 12 mm ($\frac{1}{2}$ in) steel screws, which also pass through an 800 mm (32 in) diameter 6 mm ($\frac{1}{4}$ in) thick aluminium disc and trap it between the fork and the polar shaft end. The large disc is the RA drive friction drum and is driven at its periphery by a 25 mm (1 in) stainless steel roller,

Figure 6.5 The friction RA drive.

spring-loaded in place and rotated by a stepper motor via a 500:1 reduction gear box. The reason for using a friction roller system is that I could never reduce the periodic errors of the original worm drive below about 20 arcsec and so I eventually took Ron Arbour's advice and tried friction driving instead. This is much better, and the short-term errors are now of the order of 5–10 arcsec. It is important that the friction drum is truly concentric with the polar axis, and I achieved this by bolting the drum and fork in place without the telescope and then running a Cintride tungsten carbide flexible disc against the drum periphery, while the drum was allowed to spin freely. An electric drill, clamped in a heavy machine vice, provided the motive power for the grinding disc, and gave a smooth and concentric finish (and a lot of aluminium powder!) after about 30 minutes of running under light pressure. A quartz-crystal-controlled oscillator unit feeds the RA drive motor, and allows its speed to be set to an accuracy of 0.002%.

I bolted the steel base plate to the top of a 500 mm (20 in) square concrete block column, which supports the telescope assembly approximately 2.3 metres (90 in) above ground level, as measured to the declination axis. This puts the declination axis close to the geometric centre of the 3.9 metre (13 ft) dome, which has a bearing ring at the same height as the axis, and gives the best possible clearance between the moving dome and telescope. It is, of course, necessary to prevent warm air, from the workshop below, from rising through the base frame and circulating past the telescope optics. I avoided this by using a "skirt" of bubble-wrap insulating blanket, which is securely attached to the observatory floor at the top end and to the concrete pillar at the bottom end. A protective plywood box is attached to the underside of the floor and surrounds the skirt, but does not make contact with the pillar, so as to avoid the transmission of vibration from the observatory floor to the telescope mounting.

Collimation

Any tilted-component telescope is critically dependent on the proper alignment of the optical system. However, this is also true of the Newtonian reflector, and setting up the Schiefspiegler is not really so very difficult. The situation is helped by the relative insensitivity of the position and angle of the tertiary mirror, the aberration of which is fairly slight and is easily balanced by small changes in the primary mirror tilt. If you set the tertiary at the theoretical geometric location, then it will probably not be necessary to make any further adjustments to it when setting up the system.

The procedure which I use is to locate the optical axis with an accurately positioned pointed stick attached to the secondary support (measured from the secondary mirror edge to the theoretical centre of the incoming beam). The flat is then tilted until it shows a centralised view of the tertiary mirror in the eyepiece draw-tube and the tertiary is then adjusted to see the secondary centrally. The secondary is then similarly adjusted to show the primary mirror centrally in its reflection.

Once the smaller mirrors are all in alignment, I tilt the primary until the tip of the optical axis marker stick is imaged in its centre and all the mirror discs appear coaxial in the view through the draw-tube. At this point, the optics are correctly aligned according to the telescope design parameters, and a star image should look reasonably circular when viewed through the system. However, the collimation of the Schiefspiegler is very sensitive to the primary mirror tilt and is unlikely to be accurate after the initial basic setting up. The final adjustment must be done on a star image and is largely a case of tilting the primary until no astigmatism is seen.

I have found that the collimation is no less stable than that of a typical Newtonian, and it rarely needs more than a slight tweak of the primary mirror tilt to return the images to perfection. At such a long focal length, the adjustment can be done with precision, as even a trace of aberration is easily detected.

The Results

The Schiefspiegler has been my main telescope for almost 10 years, and this speaks for itself as far as the image quality is concerned. My only desire is to construct a larger one, when an off-axis design is developed which can be scaled to greater than 320 mm aperture! I did spend some time with ray-tracing software, trying to refine the design, but it seems that a

Figure 6.6 Images of Jupiter and Saturn taken through the 318 mm f/20 Schiefspeigler with an SX CCD camera.

greater aperture is not possible within the confines of my observatory dome. This may change with the development of multi-pass designs, like that of Erwin Herrig (*Sky & Telescope,* November 1997).

There is no doubt that this instrument can define planetary detail better than any that I have used in the past, but the greatest problem is that of British "seeing"! It is almost certainly true that a good-quality Newtonian will be able to match the Schiefspiegler for resolution, except on the most exceptional of nights. Still, the telescope does work very well, and I would be unlikely to get better images from any other design of this aperture; some selected examples are shown above. I can strongly recommend the Buchroeder Quad-Schiefspiegler to anyone interested in planetary observation or imaging.

Chapter 7
A Wright Camera

Bratislav Curcic

The few amateur Schmidt-type astrographs which have been commercially available over the last few decades have suffered from the inconvenience of a curved focal surface and, because this surface is at prime focus, it does not afford a visual preview. This 146 mm (5.75-inch) f/3.9 Wright has several advantages over a Schmidt. It has flat focal plane and is significantly shorter (roughly half the size of an equivalent Schmidt). In its Newtonian form it has an easily accessible focal plane and hence can be used with ordinary SLR cameras. This makes it an ideal project for an ATM seeking to make an astrophotographic instrument.

Why Wright?

For today's ATM wanting an astrographic instrument, there are quite a few choices available: Schmidt camera, Maksutov and its derivatives, Houghton family of telescopes (including the very desirable Lurie variant), Concentric Schmidt–Cassegrain family. These are just a few astrocameras that offer exceptional image quality over a large area. So why choose a Wright? The Wright has several inherent advantages. It has a flat (well, *very* long-radius) focal plane. It is quite short (roughly half the size of an equivalent Schmidt) and, in its Newtonian form, it has an easily accessible focal plane so, combined with the previously mentioned field flatness, it can be used with ordinary SLR cameras. It also is a relatively simple design, having only two elements, albeit at a price of having them strongly aspheric. Its corrector can be quite thin, requiring little of the expensive (and difficult to find) optical glass. Its main disadvantage is astigmatism. This will limit the design to about f/4, and even then we will have to accept less than perfect images in the very corners of the film frame. But as with everything else in telescope making (and life as well!), there are compromises we have to make. Difficulties in execution, collimation, use and setup have to be weighted against each other. Wright may have the edge in some areas (relatively simple layout, very easy to use), while other designs may be much better choices from another point of view (no compromise in image quality, for example). The choice for me was clear – I didn't like the idea of handling film cuts (hypered film is *extremely* sensitive to fingerprints!) and fiddling with the curved focal surface that comes with Schmidt. I also didn't want to have more than one refractive element, because there was no anti-reflection coating service available to me at the time, so the Lurie was out too. The concentric Schmidt–Cassegrain is a great performer, but a bit impractical in small sizes. A Maksutov would still suffer from field curvature (more elements again). So the answer was the Wright.

The Theory

The Wright telescope consists of an aspheric corrector plate (a "Schmidt" plate, as it is often called, because it belongs to the family of correctors that were invented by Bernhardt Schmidt (Schmidt, 1953) and an oblate ellipsoid[1] for the primary mirror.

In the original design given by Wright (Wright, 1937) the primary's amount of deformation is exactly the same as in an equivalent parabola (but opposite in sign, of course). The corrector is also proposed to be placed at primary mirror's focus. Such a system is aplanatic (free of spherical aberration and coma), and suffers from a moderate amount of astigmatism. The corrector is of course a standard Schmidt shape, which can be approximated by

$$z = P(Ah^2 + Bh^4 + Ch^6 + \ldots)$$

where z is surface height (profile), P is the "power" of the plate (1 being just enough to correct spherical aber-

[1] The oblate ellipsoid has foci that are getting progressively shorter as we move towards the outer zones, unlike the more familiar prolate ellipsoid, whose outer zones have a longer focus, similar to a parabola.

ration of the simple spherical mirror, with 2 required for a traditional Wright), $h = \sqrt{x^2 + y^2}$ represents radial distance from corrector's centre. We can ignore higher-order elements in all but ultra-fast systems (that is, coefficients *C* and above are taken as zero). *A* and *B* are coefficients often given in literature as:

$$A = \frac{kD^2}{16(n-1)R^3} \qquad B = -\frac{1}{4(n-1)R^3}$$

where *D* is the corrector's diameter, *R* is the radius of the element to be corrected (spherical mirror in Schmidt, or paraxial radius of the oblate in Wright), *n* is the mean refractive index of the glass used for the corrector and *k* is a coefficient defining the position of the neutral zone. (The neutral zone is a part of the corrector that has no optical power.) The commonly taken value of *k* of 1.5 will give us the corrector with the neutral zone at 86.6% of the entrance pupil. Such a design has the least amount of chromatic aberration and is the one most often illustrated in the literature. As it turns out, chromatic aberration in relatively weak designs (f/4 and slower) is not much of a problem at all. Much more worrying for a telescope maker is a steeply turned-up edge beyond the 87% zone – quite difficult to make smooth. But we can tackle the problem from the other side. The neutral zone can in theory be placed anywhere from 0% (dead centre) to 100% (right at the edge), by varying *k* (Figure 7.1).

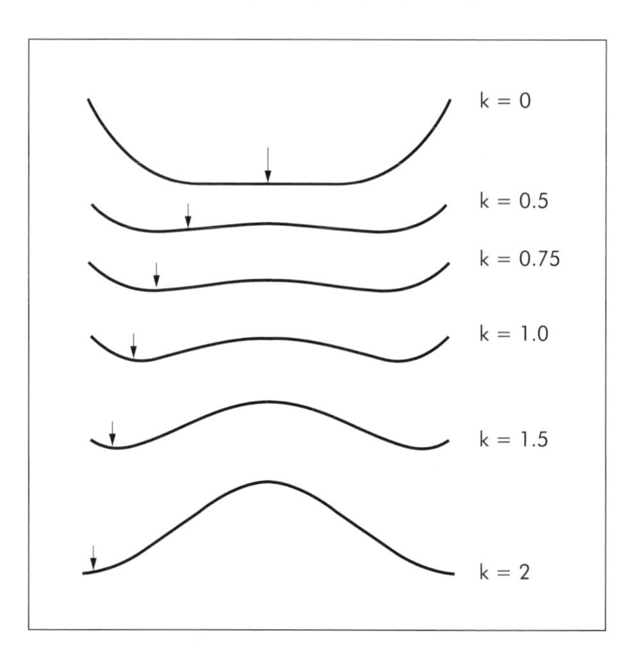

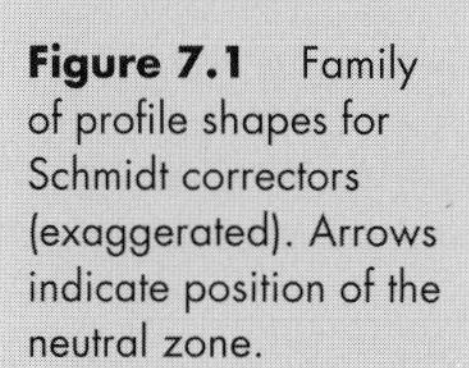

Figure 7.1 Family of profile shapes for Schmidt correctors (exaggerated). Arrows indicate position of the neutral zone.

If we choose a corrector with the neutral zone at around 70% ($k = 1$), we'll get into quite familiar waters for an ATM, as many optical elements are figured by working at 70% zone. Thus the chosen corrector is also the closest to flat (it has the identical height of central and edge zones). This has another advantage – we could use only one tool to grind both sides of the corrector initially flat.

The practical execution of the Wright telescope may or may not depart from the theory. One thing that is quite obvious from looking at the layout of the telescope as given originally by Wright is that it is impractical to convert it into a Newtonian variant, as the secondary will be quite far from the corrector plate (Figure 7.2).

This may not be such a problem after all, as we can simply support the secondary using a spider. But if we want to avoid the diffraction that comes with that solution, and decide to mount the secondary on a corrector, the supporting structure may be so long that vibrations and flexure become a real problem. Moving the corrector plate towards the primary gives us an ideal position for secondary support, but at a price: our telescope will no longer be a true aplanat. This may sound severe, but in practice the corrector placement is not nearly as critical as in some other systems. Already having some astigmatism to start with, to introduce an insignificant amount of coma is a small price to pay to have a mechanically much more sound telescope. The change in field curvature and residual coma can be negated somewhat by making the primary oblate a bit stronger, and figuring the corrector to match. A good ray trace program is an indispensable tool for optimising the system for smallest

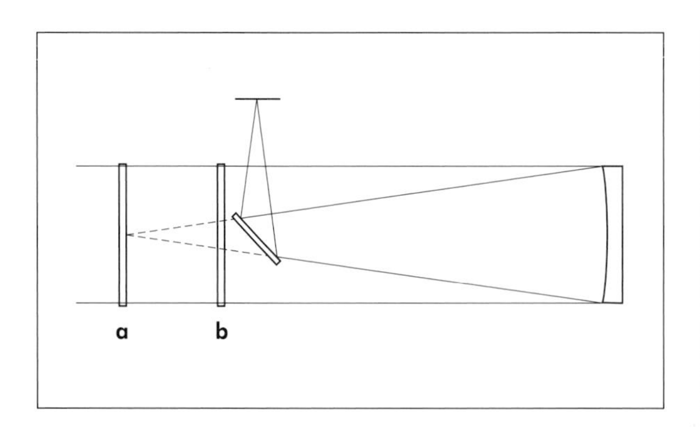

Figure 7.2 Corrector plate positioning in Wright telescopes: (a) 'true' Wright, (b) proposed position.

image size (balancing residual coma, astigmatism and curvature of the field), but in reality differences will be quite small even if we stick to Wright's original formula and just slightly respace the corrector. In any case, all necessary design details and Seidel coefficients[2] for a Wright camera are given in the literature (Rutten and van Venrooij, 1988).

Table 7.1 Full specifications for my Wright camera

*D*1 corrector diameter	146 mm
*R*1 radius of curvature	flat
*T*1 axial distance	6.2 mm
*M*1 medium	517 642
Aspheric coefficient *A*2	$4.425 \cdot 10^{-6}$
*B*2	$8.30546 \cdot 10^{-10}$
*T*2	444 mm
*M*2	air
*D*3 primary diameter	180 mm
*R*3 primary's paraxial radius	–1130 mm
Deformation of primary	1.451
*T*3 distance to secondary	390.2 mm
*D*4 secondary diameter	55 mm
*T*4 BFL (secondary to focus)	173.37 mm
Effective focal length	563.57 mm
Geometric *f*-ratio	3.86

Note: This table contains the data for a Wright as I ended up making; it is not an optimal design (for example, primary deformation of 1.55 would give better results), but after checking ray trace results and managing to get very smooth primary I decided to stop at deformation of 1.45. Coefficients *A*2 and *B*2 are 'reverse-engineered' to zero on-axis spherical aberration.

[2] The Seidel method, derived in nineteenth century, treats every monochromatic aberration (spherical, coma, astigmatism, curvature of field and distortion) as a separate calculation. The Seidel coefficients are calculated for each surface, making it immediately obvious which surface contributes to which of the aberrations and by how much. This calculation is only approximate, but it is still very useful as first iteration, before the system is optimised using the exact ray trace.

Making a Wright

After deciding on a concept, we will have to tackle a few more aspects of the practical nature if we are to have a working telescope. Before making the optics, we need to decide on a testing strategy. The primary, being a conical section, is a fairly straightforward affair. It can be successfully tested on its own (from the centre of curvature) by using any of the available (or preferred) methods : Foucault, Gaviola (caustic), Ross or Waineo null tests, and so on. The oversized mirror requirement in a photographic instrument like a Wright will make the primary even faster than the desired final f-ratio (in my case primary was f/3.2), but this is still well within the grasp of an amateur optician, especially if we remember that our goal is an astrograph, not a high-resolution visual instrument. The corrector will be tested in a completed instrument, together with a fully figured primary, either in collimated configuration (using starlight or another telescope to provide parallel rays) or autocollimated (using a large flat).

For this, we need some means of rigidly supporting the optics during testing and figuring sessions. Optics can be mounted just in a simple frame, but sooner or later we'll need a proper rigid and stable tube, so we should consider making one at this stage. The tube in a photographic telescope needs to provide just one, critically important thing: it must support the optics in the same relative position regardless of external factors like gravity and temperature. A little flexure and expansion or contraction of the tube will be entirely unnoticeable in a purely visual instrument, but film is frighteningly unforgiving when it comes to these factors. Of these two, thermal properties are more difficult to battle. Many structures, including rolled sheet metal, various truss configurations, plastic and even cardboard, can be made quite sound mechanically. But as far as tackling thermal expansion, the choice is surprisingly narrow. The old, time-proven solution is to use Invar rods, but for today's ATM there is an even better solution – carbon fibre. This space age material is quite light (lighter than fibreglass, almost half of the specific gravity of aluminium), of phenomenal tensile strength (almost twice that of titanium) and possessing a thermal coefficient of very nearly zero (one-fortieth that of aluminium, less than one-tenth that of fibreglass), yet it is still relatively affordable. There are commercial vendors making tubes out of carbon fibre composite, but with little effort an ATM can make one too. Making a solid-walled tube out of carbon fibre is possible, but quite an expensive solution. What we need is some sort of core material whose only role is to keep two relatively thin layers of carbon fibre a couple of millimetres apart. A cardboard tube is an excellent choice, but if we need the exact dimension that isn't available in cardboard, other materials can be used such as some types of plastic, wood, and so on. Metal should be avoided, as its large coefficient of expansion may force core and carbon/epoxy skin to separate. There are also other issues, such as accelerated oxidation (as carbon will form an electric potential with aluminium for example), so it is best to avoid it. My solution was balsawood – a material whose strength and ultra-light weight have long been exploited in the aircraft modelling industry. Standard strips are about 100 mm (4 in) wide, and of various thicknesses. 1.6 mm ($\frac{1}{16}$ in) will be enough for a small instrument (less than 500 mm – 20 in long). For larger telescopes, greater thickness should be considered. Making a tube out of balsa strips needs some sort of a mandrel to give support for the structure (Figure 7.3). A set of cardboard cutouts will work well.

The carbon cloth is glued to the core material using epoxy resin. As with all composites, the direction of the

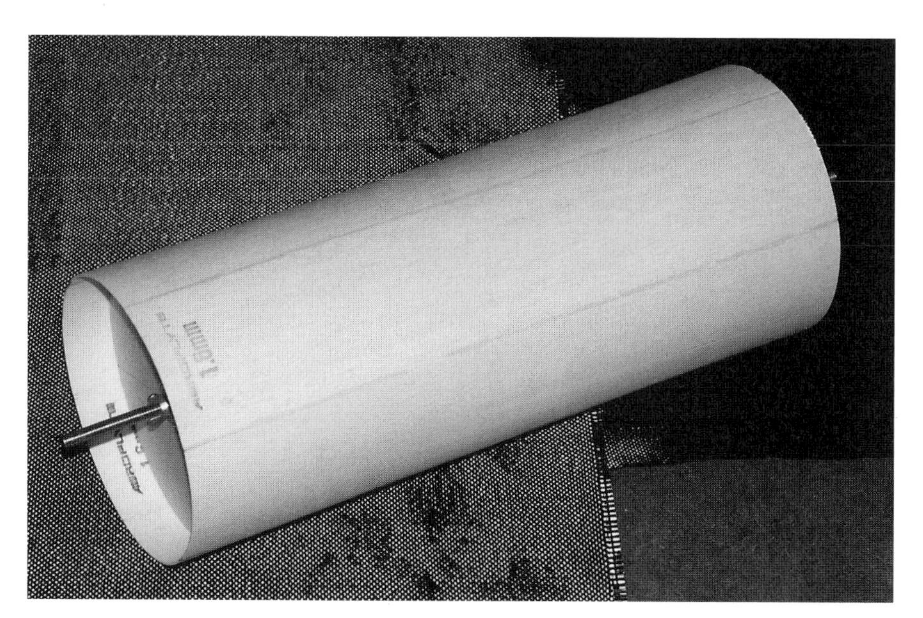

Figure 7.3 Balsawood mandrel with carbon in the background.

fibres and percentage of the epoxy are very important. Vacuum bagging is a professional method of keeping the amount of epoxy in the composite to a minimum, but just using a soft plastic spatula and squeezing and wiping off the excess resin will be good enough. We may not quite end up with a structure that has a zero coefficient of thermal expansion, but it will nevertheless have a good deal more thermal stability than almost anything else, and be extremely lightweight and strong as a bonus.

In order to keep the film plane as low as possible I took the advantage of the fact that corrector is smaller than a primary, and mounted the focuser below the outer surface of the tube (Figure 7.4). This way we do not loose any light by vignetting, yet film can be as close as possible to the secondary, minimising the obstruction and light loss.

The material for a corrector plate can be any glass that is transparent and homogeneous (free of striae and stress). Well annealed optical glass is of course most desirable, but is also the most expensive option. Clear plate ("water-white" or "crystalline" as it is sometimes known) is a very good choice, but even ordinary plate glass will work quite well. Let's remember that a large portion of commercial catadioptric telescopes out in the field use it, and that even the venerable Palomar Schmidt has had its corrector plate made of plate glass. A quick check using crossed polarisers will reveal any problems with the glass.

The primary can also be made out of plate glass, but for larger telescopes (200 mm or 8 in and above) it

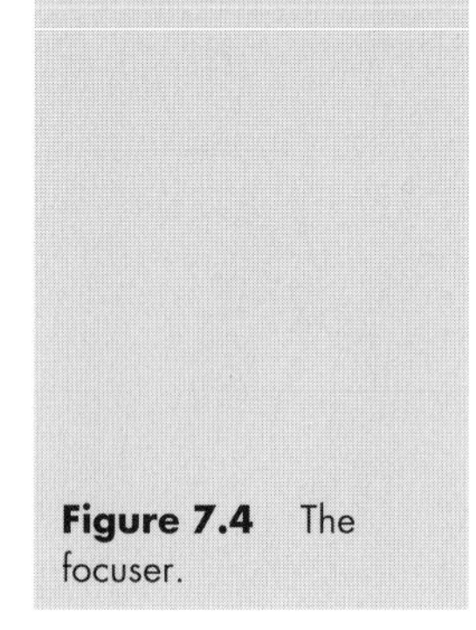

Figure 7.4 The focuser.

would be wise to use a low-expansion glass for the primary mirror, as shift in focus during long exposures may become a problem.

A simple calculation shows us that the corrector's maximal departure from flat at the 70% zone for amateur sized instruments is anything between 3 and 10 micrometres (less than 6 micrometres in my case). This is not such a large amount that it requires grinding the curve – it can be quite easily polished in from the initially flat surface. This is also the method often recommended in literature (De Vany, 1981 and 1985).

Let's now calculate the maximum allowable wedge on the corrector. The dispersion for small angle prisms is given as:

$$d = (nF - nC)A$$

where d is length of spectra in angular terms, A is prism angle, and $(nF - nC)$ is mean dispersion, often given in glass catalogues. The last quantity amounts to about 0.008 for common crown glass, and this value can be safely used for plate glass as well.

From the above formula we then get

$$A \leq 125d$$

Again, it is important to remember that the telescope we are making is a photographic instrument. There are only a handful of sites on the planet where the images of stars during long exposures do not exceed 1 arcsec. Most sites available to amateurs will be deemed excellent if long-term seeing gets close to 2 arcsec. So we can quite safely make a conscientious decision that images of 1 arcsec are quite acceptable. Hence $A \leq 125$ arcsec, which corresponds to almost 0.1 mm (0.004 in) of wedge for a corrector size of 150 mm (6 in). I set the limit arbitrarily to 30 microns (0.0012 in), playing it safe, and this amount of wedge is really easy to keep below during grinding stages. (Note: this approach is valid only for telescopes that are intended to be used as photographic instruments. For critical visual instruments, it is recommended to limit d to about one-tenth of the diffraction disc size for a given instrument (Texereau, 1984)).

As mentioned before, just one tool is required for grinding the corrector. It may be a standard glass tool, or it can be a plaster-backed or plywood-backed tile tool. A good flat cast-iron or aluminium tool will also work well. I made my tool out of plaster of Paris, covered with small ceramic tiles. For grinding of the corrector I used the standard technique for making flats. Marking two sides of the corrector I tried to spend equal time with the tool on the top and bottom of each side. That is, I ground one wet with the tool on top of side 1, the next wet with side 1 on top of the tool, then the tool on top of side 2, side 2 on top, and so on. During grinding stages, we want to get our corrector to be as close to plano-parallel – that is, no wedge and both sides flat. Wedge is checked using standard methods found in the literature (Texereau, 1984). As previously mentioned, the requirement on wedge is not particularly critical. What happens if the corrector sides don't end up perfectly flat? The consequence of having some sphericity on the corrector is almost insignificant – the spherical aberration of such a lens (if the corrector turned to be a weak double convex, for example, instead of having both sides flat) is so small that the only effect we will ever notice is slight refocusing. Nevertheless, a good spherometer is a useful tool, and it is good practice to check the radii during fine grinding and not let them depart too far from flat. Using a common micrometer and a homemade spherometer, it was quite easy to keep corrector's wedge to below 10 microns (0.0004 in) and sagitta below 3 microns (0.000 12 in) when I finished fine grinding. There is no doubt that this can easily be bettered by a careful ATM, but again it is not really necessary.

In order to support the secondary we need a hole in the corrector. Cutting it differs little from cutting the hole in a Cassegrain primary (Texereau, 1984). I simply used a slotted piece of brass pipe that was a snug fit in one of hole saws I had at hand. On a drill stand, using no. 120 carborundum and a little force, trepanning went quite fast. I decided to drill from both sides and leave a little "bridge" in the middle. This way I minimised the possibility of chipping during the exit cut, and avoided the messy gluing and sealing procedure. One note of warning though: make sure that the bevel on both sides of the hole is wide enough, or chipping may occur at a later stages of grinding (or even worse, during polishing). This operation is best done after rough grinding, as any scratches or chipping will be removed by further stages of grinding.

Polishing the corrector differs very little from polishing the mirror. The only difference is that the corrector has two sides, and one side needs to be constantly protected while the other is worked on. I used a self-adhesive plastic sheet, as used for protecting book covers and so forth. In order to avoid surface warping because of the Twyman effect[3], I polished both sides more or less simultaneously. In any case, it is advisable to polish one side to at least 25% before polishing the other side completely (De Vany, 1981; Texereau, 1984).

For a polishing substrate, pitch is the amateur's traditional workhorse, but more modern materials like ophthalmic polishing pads (Knott, 1992) can be very effective, especially on optical glass (for which they are designed). These pads do produce a somewhat inferior surface compared with that from pitch lap, but on the refractive optics (as in case of our corrector plate) roughness is much less pronounced, as surface errors are roughly halved at the wavefront. For a photographic instrument the resulting surface roughness is entirely adequate.

With polishing pads, only 10 minutes of slow, steady polish is all we need on each side to clear the corrector glass enough for the initial testing. I used quite a crude arrangement to fix the corrector in front of my 200 mm (8 in) Newtonian: four bits of eraser rubber were slotted and pegged to the spider vanes, and the corrector was simply rested against these. Needless to say, our reference bright star had better be high in the sky! Using a Ronchi screen, I confirmed that the corrector was deforming the wavefront very little – the lines stayed quite straight all the way to the very edge.

Another note of warning here: unlike pitch, polishing pads will not flow; they will just follow the surface they are glued onto. In the case of a tiled plaster tool this surface could be surprisingly dynamic. It may change with time, temperature or moisture. Some types of plaster ("dentacal" or "hydrocal") are very stable, but the plaster of Paris that I used definitely was not. For some reason, my tool warped and wreaked havoc as the polishing pads happily dug into the glass, causing all sorts of shapes, none of which was the

---

[3] An optical surface that is ground will contain a certain amount of so-called compression forces that will be relieved by polishing. This effect, discovered by Twyman, must be taken into account whenever there is a transition from a ground to a polished state, as changes in surface tension forces will cause an optical surface to warp.

desired Schmidt type! In any case, practice makes perfection and, apart from bruised ego and some extra time, regrinding and consequently repolishing the glass (even several times) will not affect the performance at all. In the end, I solved the problem by finishing the polishing within 24 hours of the last stage of fine grinding. For the next project I'll make sure that substrate is more stable, or use pitch as a backing for the pads. In this case, after removing the pads, the surface can be polished with pitch for a completely ripple-free surface.

After getting the window reasonably polished on both sides, I set up the testing apparatus. My trusty 8 in Newtonian was producing the parallel beam by placing a light source at its focus (a pinhole, or in my case a slit from a Foucault tester). To confirm that the source was exactly in focus (that is, that rays emerging from the telescope were essentially parallel), I used my SLR camera with a telephoto lens set at infinity. When I was looking directly at primary mirror, the enlarged image of a slit was very easy to focus onto. Just remember not to focus by turning the telephoto, but by moving the slit or pinhole at telescope's focus. A finder or a set of binoculars pre-focused to infinity would be just as helpful.

The Wright telescope is then aligned in front of the Newtonian and the correction checked using a Ronchi screen. A strongly oblate primary and a corrector which is essentially just a plane window at this stage produces a set of lines which are quite curved even when the screen is far from focus.

For figuring, I cut some petal-shaped pads (with the maximum area at the 70% zone, diminishing to zero at the centre and at the edge) and glued them onto the tool (Figure 7.5).

Further polishing was mainly done using normal 1/4 to 1/3 W strokes, with moderate offset. As the correction settles in, Ronchi lines will straighten. This may not happen always as we wish – polishing is influenced by many parameters, not always obvious, and not always predictable. Refractive surfaces are a bit difficult even for a seasoned mirror maker to understand. Many times I found myself puzzled by the figure, even though I did have previous experience with lenses. The golden rule is not to rush things. Think it over and, if still not sure, make a simple test: touch the glass surface for about 15 to 20 seconds near the defect that you are about to attack. Warm fingers will cause

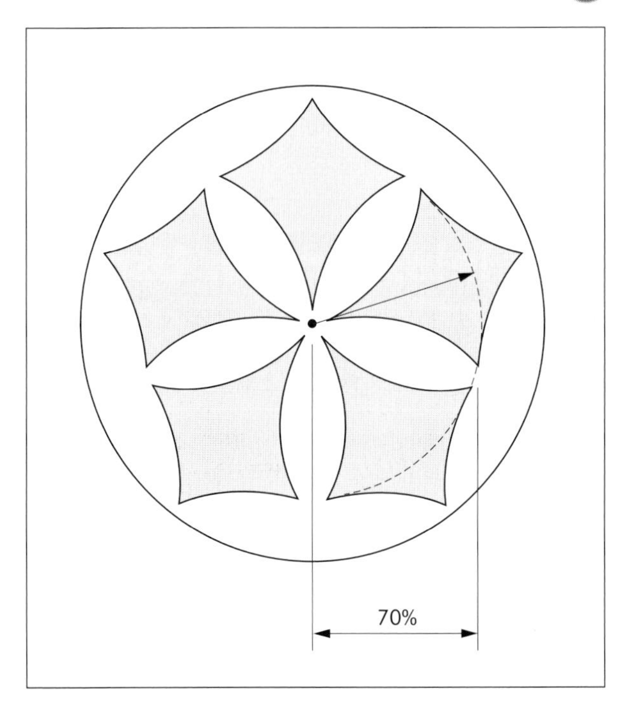

Figure 7.5 Lap for figuring a Schmidt corrector plate (dark areas are pitch or polishing pads).

glass to swell. If the defect looks less pronounced, it is a groove. If it is even more visible, it is a hill. One thing is for sure – producing a dreaded turned-down edge (TDE) isn't reserved just for making mirrors. In a corrector whose very edge must be steeply turned up, it is even more difficult to battle the familiar enemy. The advice often found in the literature, to choose a glass piece larger than needed (in order to have some room to mask it down later), speaks volumes about the potential problems. Otherwise, working on a refractive optic differs little from working on a mirror. At times it seems that figuring action is slower than on a mirror, but we have to remember that one needs to remove almost four times more glass from a lens to have an equivalent effect on the wavefront.

As with the any other optical work, there are no fast and hard rules. What works fine for someone may be disastrous for someone else. Of all methods, working with local polishers seems the most appealing, but it also gives the poorest results (roughest surface). In any case, all methods used for mirror figuring outlined in the literature will work just as well on a corrector. Another warning I can give here is that optical glass is far more thermally sensitive than Pyrex, and even plate glass. Handling of the working piece by warm hands will cause local expansion; combining this with the ultra-fast action of polishing pads on optical glass can be deadly. This effect should be taken into account most seriously.

The Telescope in Use: What One Can Expect from a Wright

After finishing the telescope (Figure 7.6), the most natural way to test it is to stick in a high-power eyepiece

Figure 7.6 Finished Wright scope in action on an equatorial with a guidescope.

and check the extra-focal images of a bright star. When I did that, I was frankly somewhat disappointed. In fact, I never made a telescope that performed visually so poorly. The residual spherical aberration was much worse than the commonly accepted quarter wave. Ronchi lines were visibly bent, and diffraction rings were deeply buried in a bright halo. Visually, I would classify my Wright telescope as a pretty mediocre one. It never really snaps into focus. It shows bands on Jupiter, but little else. It splits relatively close doubles, but not very convincingly at all. It can also resolve bright globulars (ω Centauri and 47 Tucanae) into thousands of stars, but quick a glance through the real thing (my 8 in Newtonian stopped down to the same aperture) simply demolishes it.

On the other hand, this is a *photographic* telescope, and I had to remind myself about that. Even at almost twice the size of the theoretical diffraction disc (at about 8 microns instead of theoretical 5 microns for an f/3.85 instrument), images are still smaller than seeing will allow on even the sharpest emulsions. Test exposures showed tight and round images. A 10 minutes exposure of ω Centauri was swarming with hundreds of tiny specks, limited only by the resolution of the film (Figure 7.7). On an image of the Milky Way area, stars in the corners did show some softening, as expected, but still looked quite good. Any star brighter than just

Figure 7.7 Enlarged photograph of ω Centauri, demonstrating on-axis resolution.

above detection level would swell enough to cover up the slight elongation (slightly asymmetric images in the very corner are combinations of astigmatism, vignetting and a little residual coma). Initial tests also showed that something wasn't quite square, because some corners looked better than the others (the secondary placement in a system like this is very critical). Collimation of such an instrument is not a trivial affair. I will only say that after spending several nights collimating using the eyeballing technique, sight tube, Cheshire and other common tools, I decided to buy a laser collimator. I'm not saying that it would be impossible to collimate without such an instrument, but it certainly makes a difficult task much easier.

Can an instrument as complex as a Wright really be made to the highest visual standards? I'll say that the answer to this is a resounding "Yes!", but it would definitely be not an easy task. As it is now, I'm not too eager to try to improve the correction to "diffraction-limited" standards. With anything but the finest-grained emulsions and absolutely the best nights (perfect guiding too!), there would be no visible improvement at all. Besides, improving the correction in the centre will not improve off-axis images, which are already several times larger than images in the centre of the field. Most of my eyepieces also have really tough times with a steep f/3.8 cone, so even visually there is little gain to be had. The eyepiece's astigmatism is covering the telescope's aberrations everywhere except in the very narrow area in the centre. Well, maybe one day ...

Figure 7.8
5× enlargement of a corner of an image of the Pleiades, showing off-axis performance of a Wright.

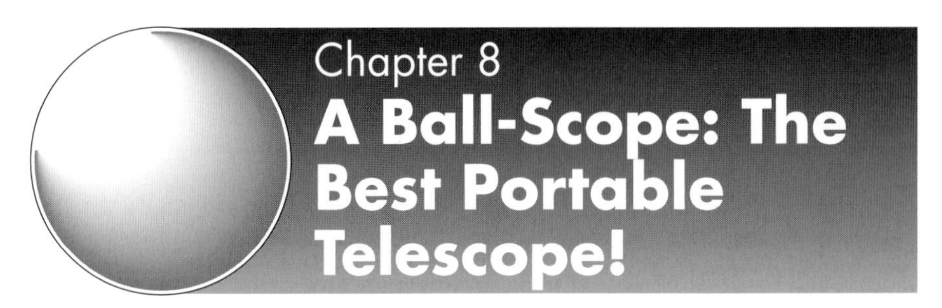

Chapter 8 A Ball-Scope: The Best Portable Telescope!

Steven Lee

Equatorial mounts have a "difficult spot" near the pole; altazimuths have it near the zenith. This ball-mounted reflector avoids both problems. Perhaps the best-known example of this type of mounting is the Edmunds Astroscan. Steven Lee has successfully extended this principle to a 12.5 in (315 mm) telescope, showing, on the way, how a typical ATM overcomes the problems that beset him.

I am in the fortunate situation of working and living far away from any large city, and so I am able to pursue my hobby from home, completely unhindered by light pollution and without any need to travel to find dark skies. This affects the way I build telescopes and the emphasis I place on their design. However, I was given a challenge to build a telescope which would best suit the needs of amateur astronomers who live in cities and so must have portable equipment. After a lot of thought I settled on the following design, which I think is one of the most useful designs for a medium-sized portable telescope for the "average" observer who wants the satisfaction of building his or her own scope. The design is a little unusual and more difficult to make than the standard Dobsonian design (which is the most common type in this range of telescopes), but the effort is well worth it. I have to admit that the design is not original, and you can even buy several commercial versions of this type, one of which is excellent (and very similar to the one presented here, for reasons which are obvious – we both had the same design goals and ended up with the same solution).

Design Features

As I see it, the Dobsonian design is popular as it allows one to cheaply mount a large set of optics and make them transportable. As such they work quite well, but whenever I use one (especially a poorly designed one) I note a couple of shortcomings. The most significant is the uneven force necessary to push them around in each axis, most noticeably the azimuth axis when the telescope is pointed near to the zenith, making it difficult to use in the best part of the sky. This feature is disparagingly called "Dobson's Hole" by some. Another aspect is that often the configuration is not all that compact; such telescopes frequently have solid tubes (usually of cardboard) and hence are large and heavy. A better approach is to dispense with the solid tube and use a truss arrangement, but many medium-sized commercial telescopes of this form are supplied with solid tubes.

What I was looking for was the *best* design for a medium-sized portable telescope. By medium-sized I mean one that is large enough to produce bright views of deep-sky objects and yet still be easily transported and erected by a single person. To me, this means a mirror diameter of somewhere between 25 cm and 36 cm (10 to 14 in), depending on whether you prefer more aperture or better portability. I eventually decided on a "ball-scope" – a Newtonian telescope mounted in a spherical shell and supported on a simple three-point base. The telescope is pushed around just like a Dobsonian but has several advantages over that design:

- The ball-scope does away with Dobson's Hole because it has no single axis, and the force to move the telescope is always the same no matter where you point it.
- Because the mounting has no definite axis, the assembly can be rotated to bring the eyepiece to the most comfortable position for viewing.
- The mounting and mirror box – two separate components in a Dobsonian – are one unit in a ball-scope, making for a more compact arrangement when transporting it.
- If designed properly, the top end can be stowed inside the ball for an even more compact arrangement.

The operating principal of a ball-scope is the large truncated sphere that acts as mirror cell, part of the tube and mounting all in one. The centre of gravity (CG) must be at the centre of the sphere for it to work, and this places some restrictions on the overall size of telescope that can be made this way. A quick look at Figure 8.1 will show that to balance a weight placed at the eyepiece end of the telescope, a lot more weight must be placed at the mirror end in order to satisfy the balance condition. This is where it departs from the Dobsonian design, where, as you add weight to the top you simply move the altitude bearings higher to accommodate the higher CG. No such freedom exists in a ball-scope. If you make the top too heavy, you must add weight at the bottom to compensate; so careful design work is required in order to achieve success. However, the higher the CG, the less stable the configuration, and so it does pay to have the CG as low as possible (and many Dobsonian builders place extra weights at the rear of the telescope in order to lower the CG for just this reason). The ball-scope should be a very stable system.

I had another goal in my design – a comfortable viewing position with no need for a ladder to reach the eyepiece when at the zenith. This places restrictions on the focal length of the mirror, and, as we already have a limit on its diameter from weight restrictions, the focal ratio. As this is a custom telescope, I chose the height of the eyepiece to suit my own eye height – about

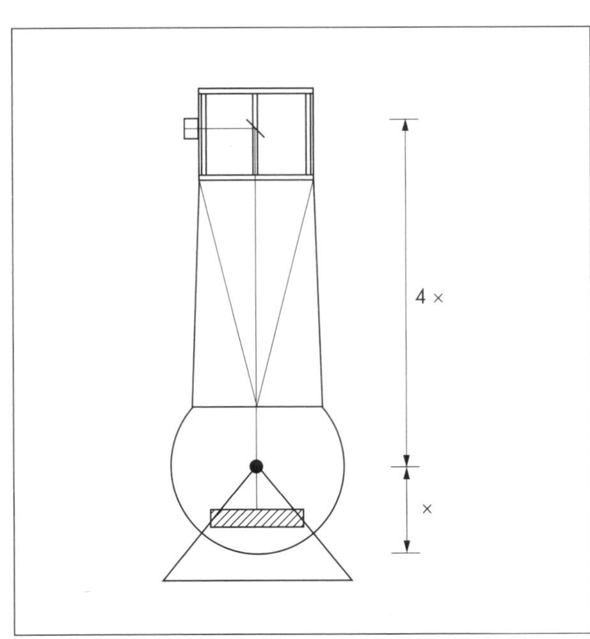

Figure 8.1 Rough layout of the ball-scope.

1.65 m (65 in). As an arbitrary starting point, I assumed that the focal length of the mirror is about how high above the ground the eyepiece will be; any differences can be accommodated by adjusting the height of the stand. (The "average" eye height would be somewhere between 1.5 and 1.8 m (59 to 71 in) – rather a large range and almost impossible to accommodate in a single design.) With the largest mirror size I would consider, namely 36 cm (14 in), this yields about an f/4.5 mirror. For visual use I don't like mirrors faster than f/5, because of the poorer performance of eyepieces with fast focal ratios, so I looked at the next smaller standard mirror size – 32 cm ($12\frac{1}{2}$ in). This works out around f/5, and seemed perfect, and is what I decided to use.

To get a better understanding of the job ahead, I made accurate drawings of the proposed telescope and made estimates of weights of materials. As I've already pointed out, for each amount of weight added at the top of the telescope much more needs to be added at the bottom in order to keep the balance at the centre of the ball – about four times as much it turns out, so using a full-thickness mirror is no disadvantage. Knowing the weight of the mirror (9 kg – 20 lb in my case) and its rough position in the ball gave some indication of how much weight can be carried on the top end. Unfortunately, my calculations showed that I would have to dispense with a heavy finder and focuser (and eyepiece!) to achieve balance without adding weight below the mirror. This wasn't acceptable as I like 50 mm (2 in) aperture finders (heavy), JMI NGF focusers (quite heavy) and my 16 mm Nagler eyepiece (amazingly heavy). So it looked like I had to compromise on either equipment or weight. The weight lost the argument, but I knew that I had to make the top end as light as possible and the bottom end weight had to be as low as possible.

There were a few other points to consider before starting construction. The top end has to fit into the ball for transportation and so this defines both the diameter of the top end and the size of the hole in the top of the ball. There are other interlinked parameters, too. The mirror cell must be designed to place the mirror as low as possible into the ball, and its height defines how high the top end can be (unless you don't mind it protruding above the ball during transport). In fact the mirror cell design is quite integral to the success of the telescope. A cover to protect the mirror must be made, and you also need to consider whether certain accessories will be built into the scope – a mirror cooling fan, for example, and battery power for it and other devices. It might prove difficult to incorporate these later. I will stress again how important the total design is – without a good overall plan of what you are making, you could be in for some surprises. This design is not one suited the slap-happy, bang-it-together-and-see-how-it-goes builder. But for those who do take pleasure from a job well done, then this design will produce the *best* portable telescope possible, I believe.

With these constraints, I set about finalising the design and then to making it.

The Ball (No. 1)

The most difficult aspect of the ball-scope is the manufacture of the ball. This is hardly surprising, as it represents such a significant fraction of the entire telescope. Other than the optics, it is the component that must be built most accurately for the telescope to function properly. There are several ways to build the ball; here are two that I have tried.

My first attempt at the ball was to get it "spun". This is a technique whereby a flat sheet of metal – usually aluminium – is pressed against a form of the desired shape while the sheet is rapidly spinning. (This is the method that is used by the commercial version of this telescope.) The technique is clearly not available to the average worker in their garage, but looking through the telephone book I found a page of workshops which advertised metal-spinning, so it is reasonably commonplace (at least in big cities). My first inquiry was successful, too, finding a place which had a hemispherical former of the desired diameter (600 mm – 24 in) and were willing to fabricate the ball at (what I thought was) a reasonable price – A$150 for the two hemispheres, and A$100 to cut the hole (450 mm or 18 in) and weld the two halves together. This gave me great hopes for quick completion, as in theory the spun ball would require only a little smoothing before being powder-coated and finished.

However, the practice did not go so well – although knowing what I know now should produce a successful outcome. The fault was that I had asked for the ball to be made from aluminium 2.5 mm (0.1 in) thick and they were reluctant to do so, convincing me that the completed shape would be sufficiently strong in material only 1 mm thick and significantly easier for them to make. As they were the experts and had made similar things before, I accepted their advice and the ball was duly manufactured. The first inkling of problems to come was when the ball was delivered to me. I thought that I had successfully conveyed my intentions with the ball, stressing smoothness of finish but when it arrived it wasn't what I had expected. The seam, instead of being flush with the surface, protruded some 20 mm ($\frac{3}{4}$ in) out of the ball – a rather significant impediment to free movement of the ball. The problem arose because I live 500 km (300 miles) from where the ball was made. Had I been able to visit the factory and show pictures of what I wanted, I'm sure the outcome would have been more successful. The ball was also significantly egg-shaped, although I didn't think that this would be too much of a problem.

The factory took the ball back and redid the seam (at no charge). Unfortunately, the newly welded ball was still significantly non-round. They said that they couldn't really do any better due to the limitations of the technique, but I suspect that it might have been a lack of experience or motivation that was the problem, as it can be done better by others. So the lesson is to visit the factory and fully discuss the project with the manager to ensure that they know what you want and so that you can see that they make exactly what you need.

None the less, the ball (better described as the "egg") was turned into a successful telescope (at least for a while). The brackets to hold the mirror cell were welded into the ball (by a local welder), and a plywood strengthening ring was attached to the top (see later). The ball required smoothing (as expected), as the welded seam, and a few other blemishes, required filling. I obtained some epoxy which is designed for filling aluminium and filled and filed until the external appearance was smooth. This process wasn't too difficult; the filler flowed quickly, so you could only work on small areas at a time, but hardened in half an hour or so and could be filed after a few hours. I spent about 2 weeks of evenings on this process before I considered its surface smooth enough to be called finished.

The ball was then powder-coated to provide a tough finish to the exterior. Powder-coating is a wonderful process whereby a special paint is applied in powder form to the surface and then baked on at high temperature. It cost me A$20 to have my "egg" placed on the production line along with other parts being done; well worth it. The filler I used actually softened and outgassed slightly when heated, which produced small bubbles in the paint. However, this never affected the smoothness of operation of the telescope but is something to check on if you follow this plan when making your own ball.

Once assembled, it needed more weight on the bottom to balance properly, so I simply melted some lead and poured it into the ball. It naturally conformed to the right shape and was as low down as I could get it. I attached a steel bar across the bottom to secure the lead once I had the right amount in place.

Figure 8.2 (*overleaf*) shows the finished "egg-scope", which worked quite well. Its non-roundness wasn't a great problem and the powder-coating proved tough enough for the job. (It should be noted that the commercial version of this scope recommends waxing the surface (with a silicon-based car polish) for reduced friction. I never found this necessary, but is something to remember just in case.) Unfortunately, the "egg" eventually failed, as the 1 mm thick skin just wasn't up to holding the weight on 3 small contact points and the pressure eventually buckled the skin in several places. Had the ball been more spherical it might have worked, as the failure points were where there was little surface curvature and hence less strength. A 2.5 mm (0.1 in) thick skin should work perfectly, provided you can find somebody to make the ball properly.

The Ball (No. 2)

Following the failure of the spun ball, I embarked on an alternative method – glass fibre and resin. This was in fact my initial plan, but I was seduced by the apparent ease of a spin-formed ball. The only drawback to this method is that it needs either a mould to build the shape within, or a form upon which to build it up. I was quoted A$2000 by a professional company to make a wooden mould (it took me no time to dismiss that plan), and so I was left to think up some cheaper means. I knew of one such ball which had been constructed using a large balloon as the form and I was

Figure 8.2 The finished "egg-scope".

intending to pursue this idea, or perhaps making the form from papier-mâché. However, these ideas were no longer necessary as I now had an almost perfect form upon which to make the ball. I split the "egg" and removed a section from its equatorial region in order to make it more spherical, rejoining the pieces with short sections of aluminium plate riveted between the two halves. I also made a template from a sheet of cardboard with a segment of the desired 600 mm (24 in) diameter to show where I needed to build up the shape.

My introduction to fibreglass and resin construction was well over 20 years ago when my elder brother and I made tubes for our first telescopes. Strictly speaking, my brother was the one who did it – I just got to watch while he had all the fun. So making the ball was really my first attempt at fibreglass construction; and if I can do it then just about anybody should be able to. The technique is relatively easy, just a bit messy and smelly.

Basically, one has woven fibreglass cloth which gives it strength, chopped matting which is used for adding bulk, and the polyester resin which holds it all together. The resin is a viscous liquid which solidifies soon after the addition of a setting agent – how long depends on the temperature and amount of hardener added – between 10 and 30 minutes is the typical setting time. The resin may be coloured by the addition of a special colouring material, so the colour permeates through the whole job.

One must first ensure that the finished product can be released from the mould or form. This is done by coating it with a release agent, or perhaps simply covering it with a sheet of plastic. In my case, I didn't actually want it to come away from the form, so I did the opposite and roughened the ball to help it stick. (This is where I found out how tough the powder-coating is – I couldn't actually remove it from the ball. I tried a wire brush, then I put a blow-torch to it; all to no avail. Once it's on, it's on.) Then it is a matter of applying a coat of resin to the surface and covering it with a layer of cloth and applying more resin. Any bumps are sanded down before the next application of resin. This process is repeated time and time again until the desired shape is achieved, the necessary thickness is attained and the finish is acceptable. I spent many weeks on this exercise, at half an hour per cycle. The covering of fibreglass varies, from a few millimetres to more than 15 mm in some places where I had to build it up a lot. Without the aluminium base, the skin would need to be at least 6 mm ($\frac{1}{4}$ in) thick to be strong enough to support the ball.

The bottom of the "egg" was considerably flatter than required, so while I was in the process of building it up with fibreglass I took the opportunity to embed a few kilograms of lead as filler. As I already knew that I needed more weight in the bottom, this gets it as low as possible for maximum advantage. I removed the lead I had already poured into the bottom, but expected to put some back later when the reworking was finished. I left three bolts protruding through the ball to help secure any extra lead added later (which I did). (I wasn't happy with the original method of securing the lead at the bottom of the ball – the bar got in the way of the fan.) If you're building a fibreglass ball from scratch, it might pay to use some lead as filler, as you're almost certain to need some.

Strengthening the Ball

I mentioned earlier the plywood ring that was placed on the top of the spun ball for added strength, but I should elaborate upon its purpose a little more. The spun ball is quite weak once a hole in its top is made – it needs to be strengthened or else it will collapse. I decided to do this by adding a ring of 12 mm ($\frac{1}{2}$ in) plywood around the top. It is 40 mm (1.6 in) wide at the top, but cut at an angle to match the curvature of the ball. The ring is attached by nine pieces of angle aluminium suitably bent to match the ball's shape (Figure 8.3). I riveted these pieces to the ball, and screwed the ring down to these brackets. Initially, I only used six brackets, spaced every 60°, straddling the mounting points for the truss (see below) but this proved too weak, allowing it to flex as the telescope was moved over the sky. The ring is strong only in the plane of the hole, the aluminium brackets providing the strength along the axis. When I made the fibreglass version, I added three extra brackets directly under the truss blocks to solve this. (I initially didn't place any of the brackets under the truss block, as I thought that any

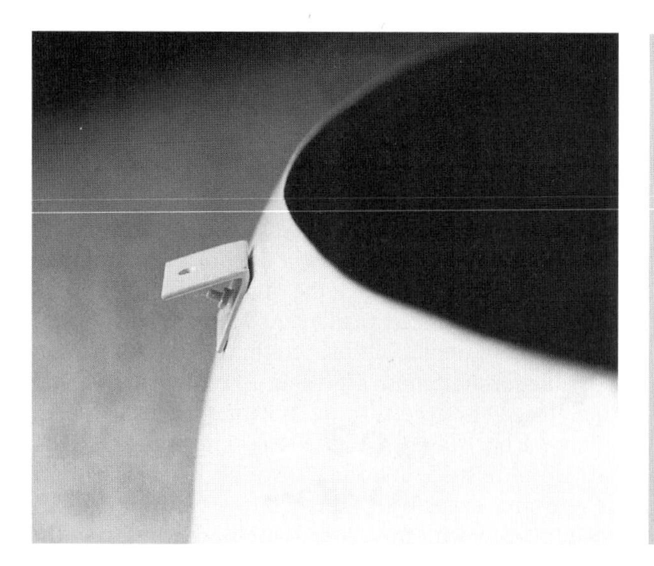

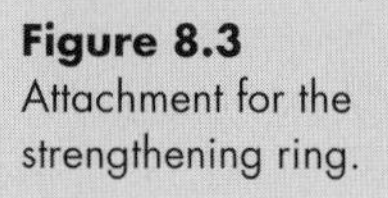

Figure 8.3 Attachment for the strengthening ring.

screw through the ring would interfere with the blocks. This proved not to be a problem once I actually did it.)

The fibreglass version of the ball retained the plywood ring, but I covered it in a layer of cloth and sealed it with resin. This increased the strength of the ring, and had the secondary benefit of hiding the gaps which existed between the ring and the ball, which showed how badly I had cut out the ring. (For those with a better tool kit than mine, the ring could be cut with a router rather than a jigsaw. This leads to a much neater hole, but means that there is slightly more wastage when cutting, owing to the larger cutter.)

There would be other ways of strengthening the ball, but I chose this method for two reasons. It provided a convenient all-round carry-handle, and it had a flat surface to which could be mounted blocks to hold the truss tubes. This shows the interlinked nature of this project – solving one problem points to the solution of another.

Stand

The ball sits on a three-point stand (Figure 8.4), floating like a Dobsonian on Teflon pads. The stand could easily be made from wood, but as it would spend its working life on damp ground I decided to make it in steel, welded together from 20 mm ($\frac{3}{4}$ in) rectangular hollow section (RHS). It should really be powder-coated for

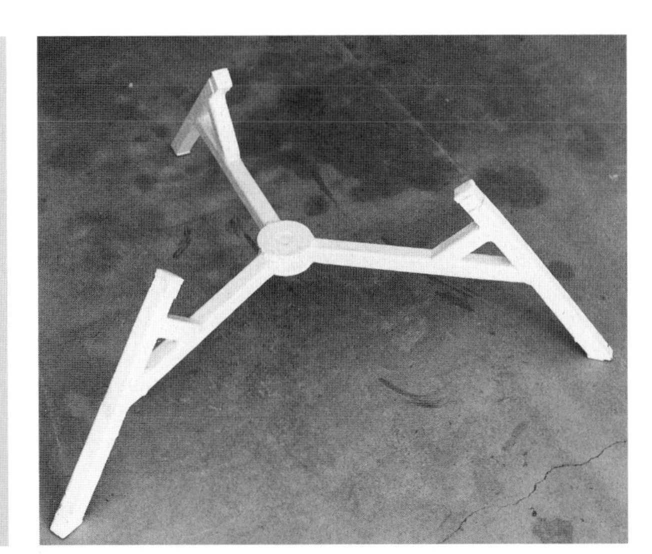

Figure 8.4 The stand on which the ball rests.

added protection, but so far I have only given it a couple of coats of paint.

The legs point in to the centre of the ball at an angle of about 45°. Much as with a Dobsonian, the positioning of the pads will modify the friction (and stability) of the telescope as it is pushed about. The height of the legs is chosen to bring the telescope to a comfortable eye height when it is pointed at the zenith. As a minimum, they should be high enough to give the ball clearance above the rough ground upon which it will normally be used. The bottoms of the legs are cut to meet the ground squarely, and should also be sealed to stop it from sinking in and then bringing the ground back when it is picked up. You can buy square plastic chair ends in most hardware shops that will do the job nicely.

You can't put the Teflon pads directly onto the steel legs (without going to a lot of bother), so I nailed them to the ends of wooden dowels. I made the dowels about 100 mm (4 in) long and trimmed them to be a snug fit into the hollow steel legs. They are just hammered into the legs, the friction being enough to hold them in place. The Teflon pads can then be nailed or screwed in place. Make sure that the heads of the nails are counter-punched below the surface so as not to scratch the ball.

Since making the stand, I have seen donuts of a Teflon-like material sold as aids for moving furniture about. I think these might prove better for floating the ball upon instead of the squares of Teflon I used, as it would allow a larger area of material to be in contact with the ball, as well as providing a better way to attach it to the stand.

Mirror Cell

While the ball was being made I started on the mirror cell. You need to think about the design of this cell in conjunction with the ball, as it plays an integral part in the whole system. The cell is very different from those made for Newtonians and Cassegrains, as there is no way to get behind it to adjust collimation. It also must be as low as possible, although for once there is every incentive to make it as heavy as possible, because this is where weight is wanted.

Initially I thought that the only way for it to work properly was to have it hang down from fixed points on the ball, with the collimation screws then pulling it up. But the obvious way of doing this would mean that it would be difficult to insert and remove. I finally worked out an offset system that loaded from the top, with the collimation bolts pushing the cell down. This would mean that I could get it down as low as was possible. It's a little difficult to visualise, but Figures 8.5 and 8.6 should help.

The basic cell is an aluminium plate 6 mm thick and approximately the same diameter as the mirror. I only cut it roughly round on a bandsaw – it is hidden and so I didn't bother making it look pretty. To properly support a mirror of 32 cm ($12\frac{1}{2}$ in) diameter without

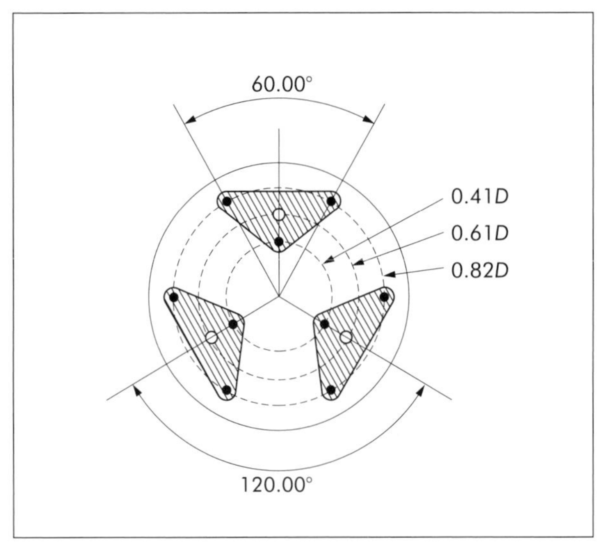

Figure 8.5 The layout of a nine-point mirror support.

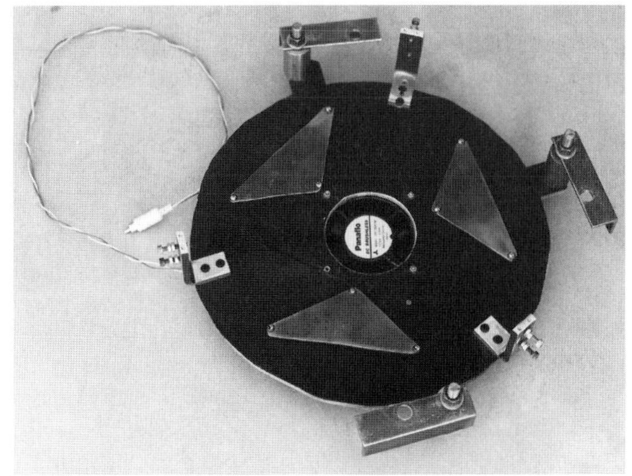

Figure 8.6 The completed mirror cell.

flexure, it should really have a nine-point support (Figure 8.5). I made mine very simply. Three triangular plates of 2 mm (0.1 in) steel were cut to size, and their apices were drilled and tapped for small, round-headed screws which would act as the supports for the mirror. Two of the three screws on each triangle are cut short so that they didn't protrude below the steel, but the third was deliberately left long. This screw passes through a suitably placed hole in the aluminium plate to stop the triangle from rotating. The hole is sufficiently large to not interfere with any up-down motion of the triangle. (If you have the ability, then removing the thread from this screw to leave a smooth rod would be better, but in practice I've never had any problems with the screw threads catching.) At the centre of each steel triangle I made a depression which is designed to sit on a round-headed screw. These screws are attached to the aluminium plate at the proper radius. Overall, the cell is extremely low and yet very rigid – a perfect combination for the purpose.

A 76 mm (3 in) diameter hole is cut through the centre of the aluminium plate for a small fan, which is used to help cool the thick mirror to ambient temperature as quickly as possible. This was cut very carefully with a large hole-saw on a slow drill press.

I believe in mirror collimation systems with axes that operate at right angles rather than the awkward 120° triaxial system usually used. This still uses three points, but arranged as a right-angle triangle rather than an equilateral; the right-angle apex is the pivot, the adjustments being done at the other two points, giving an up-down, left-right adjustment pattern.

The aluminium base-plate has three "ears" protruding from it, with holes for large bolts (I used $\frac{3}{8}$ in threaded rod because I had some on hand and it can be cut to exactly the right size). Steel posts 36 mm ($1\frac{1}{2}$ in) high (or to suit) are bolted to these lugs. From the top of each post, a 50 mm (2 in) long, 4 mm ($\frac{3}{16}$ in) thick piece of right-angled steel is attached. At the other end of this piece of steel is an oversized hole for the collimation bolt to pass through. The collimation bolt is another length of $\frac{3}{8}$ in threaded rod, with one end firmly attached to a bracket, which in turn is welded to the ball. These three brackets are made from 6 mm ($\frac{1}{4}$ in) thick pieces of right angle aluminium, with their ends shaped to fit the ball's curvature, then welded to the ball (only possible if the ball is aluminium; they would need to be an integral part of a fibreglass ball). Around each collimation bolt is a strong spring (mine are engine valve springs and are entirely adequate) which supports the mirror cell. Tightening the nuts on the collimation bolts compresses the springs and forces the cell down. If this is done right, the springs are completely compressed when the cell bottoms out on the ball. Loosening the nuts a little is all that is needed to collimate the mirror. There is plenty of force being exerted from each of the springs, so there is no chance of the mirror cell shifting as the telescope is moved over the sky.

One of the strengths (and weaknesses) of a classical Dobsonian design is that the edge of the mirror is supported in a sling. While this is fine in an optical workshop, I am always a little reluctant to trust it in the real world of pushing a telescope about. Fortunately (or not, depending on your opinion), you can't do this with a ball-scope because it can be oriented in any way to bring the eyepiece to a comfortable position, and so the mirror would fall out of a sling. To laterally support the mirror and ensure that this can't happen, you must revert to more formal restraints. Attached to the baseplate are three posts of 6 mm ($\frac{1}{4}$ in) aluminium, 20 mm ($\frac{3}{4}$ in) wide, and as high as the mirror is while resting on its nine support points. Through the side of these posts are 4 mm screws which are used to position and restrain the mirror. I use two screws per post at about one-third and two-thirds of the mirror height (Figure 8.7). You

Figure 8.7 The mirror retaining clips.

operate these screws by turning them until they just touch the mirror and then locking them in place with a pair of nuts on the outside. This way, they firmly position the mirror but don't distort it. Care must of course be taken as it *is* possible to over-tighten the screws and distort or even damage the mirror. It should be possible to actually turn the mirror in the cell with the screws tightened correctly, but only just.

On the top of the posts are 2 mm (0.1 in) aluminium plates which protrude slightly over the mirror's surface to stop it from falling out should the telescope become inverted (this is actually very hard to do because of the weight at the bottom of the ball, but I don't believe in taking risks, however small). The underneaths of these plates are covered with a "soft" material like cork so that if they do actually touch the mirror's surface it won't be damaged. The plates are screwed to the posts with 3 mm screws, with spring washers between each plate and post to allow the plate to be adjusted to the right height – which is almost but not quite touching the mirror.

All the places which come into contact with the mirror – the heads of the nine support screws and the tops of the six lateral screws – are tipped with blobs of solder. Solder is an excellent material to use as a contact point because it is soft and pliable under pressure and yet firm enough to withstand the rigours of constant use.

Mirror Cover

To protect the mirror while the telescope is not in use, a protective cover is placed over it. The cover has a secondary function, namely to hold the top end while the telescope is in storage.

I made the cover from the circle of plywood cut out from inside the strengthening ring. It is exactly the right diameter to fit through the opening of the ball, and large enough to cover the mirror. Three holes were cut so that the threaded rods which hold the mirror cell in place could pass through. So that the threads wouldn't continually abrade the holes, I glued short pieces of copper tube into the holes. The tube was large enough for the rod to pass through easily, but not so loose that it rattled about. I placed a handle in the centre so that there was something to hold while I am

pulling the cover off or putting it back on. When pushed fully down, the cover rests on the top of the mirror cell so that it can't touch the mirror.

To hold the top end in place, I mounted three pairs of wooden blocks to surround the three truss blocks on the bottom of the top end. When the top end is placed between these it can't move any way other than up. I've not found it necessary to make any more rigid restraint for the top end.

The Truss

To hold the top end to the ball, a truss is made from aluminium tubes (Figure 8.8, *overleaf*). Each of these is of approximately 20 mm ($\frac{3}{4}$ in) OD with a wall thickness of 1.5 mm (0.06 in), costing A$25 for a 6 m (20 ft) length. Six are used instead of the eight usually seen on Dobsonian systems. Dobsonians usually have a square mirror box and so it is easier to construct an eight-pole truss system than six-pole one, but six are perfectly adequate for the purpose, and are lighter too.

The most difficult aspect of truss systems is how to hold them securely in place when observing, and yet make them easy to assemble and disassemble. This is often the stumbling block of portable truss telescopes. There are many solutions to this problem, some extremely elegant. I wanted a system that was both simple to make and simple to operate, two attributes not often compatible. The one I ended up using satisfies both criteria reasonably well, although I will admit that construction isn't totally trivial if you don't have a drill press. I'm sure there are many other equally workable solutions.

The system relies on wooden blocks in which holes for the poles are cut at the appropriate angle. The blocks are then split, and a bolt passing through them used to force them together and thus clamp the poles in place. The height of the block is chosen so that a pole will sit in it without falling over when the clamp is loose, thus making for a quick and simple assembly. My blocks are about 90 mm long, 45 mm deep and 30 mm high ($3\frac{1}{2} \times 1\frac{3}{4} \times 1\frac{1}{4}$ in).

The only difficulty is in getting the angle right for the holes. The poles slant inwards towards the top end and forwards towards the next block, meaning that two angles have to be calculated. Of course, I took the easy

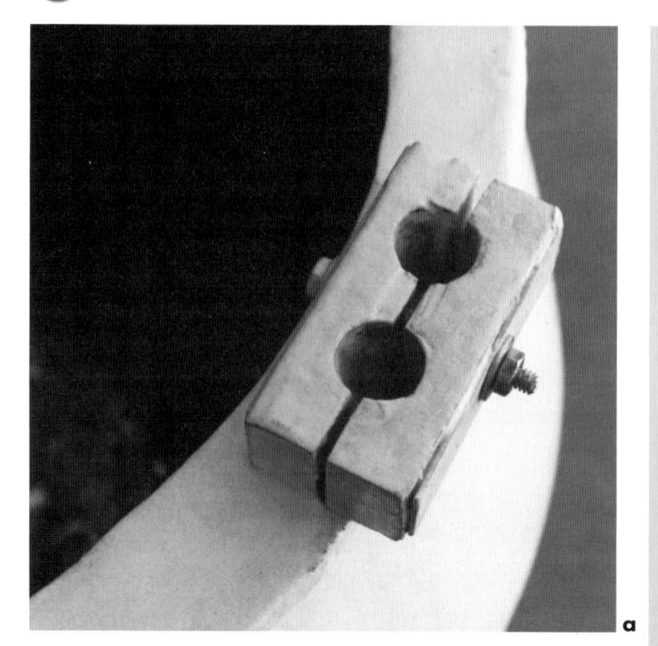

a

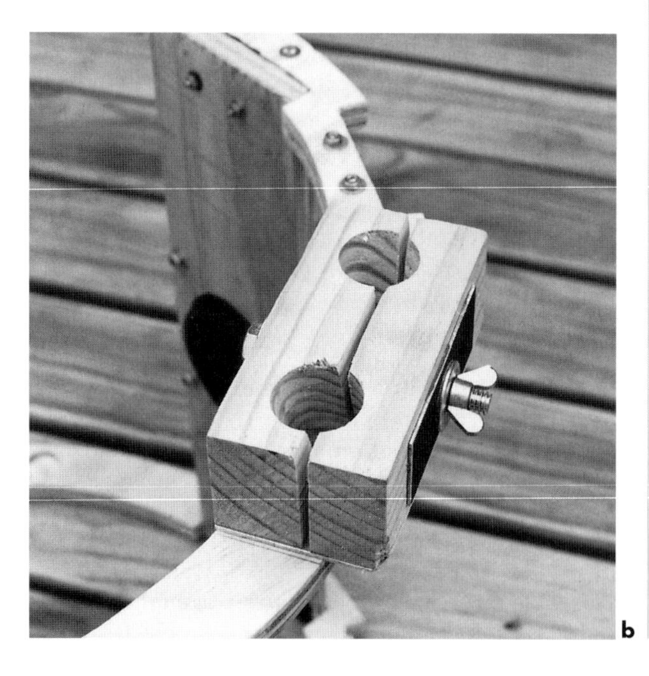

b

Figure 8.8 The truss-pole clamps on (**a**) the ball and (**b**) the top end.

way out and did it by trial and error. I set the tilt on the table of my drill press to approximately the correct angle and clamped a spare block in the vice also at approximately the correct angle and made a test hole.

After a couple of test holes I had it close enough to go ahead and make the blocks. It is obvious that the pairs of holes in each block are different, so you do one hole in each of the blocks then change the angles and do the other holes.

Once these holes are drilled, you then put a hole for a bolt to pass all the way through. I used a $\frac{1}{4}$ in bolt. The blocks are then split longitudinally. The blade-width cut away here ensures that when you tighten the bolt, the poles will be gripped before the blocks close. To ensure that the force the bolt applies is spread evenly over the block, I glued 1 mm aluminium plates across each face of the block. I finally glued the bolt into one block so that a tool wouldn't be needed to hold it during assembly. I used a good, long-curing, two-part epoxy for this.

Where the truss poles rest against either the top end or the strengthening ring I placed another small piece of 1 mm aluminium to act as a reference surface. If the lengths of the truss poles are all identical, then perfect collimation will result at each assembly. A hacksaw and file are adequate for making the poles the same length, but cutting them using a jig ensures a perfect match.

I purposely cut the poles about 25 mm (1 in) longer than I had calculated on the assumption that if I had made a mistake I can always cut them shorter, but I can't make them longer (without buying more). When I finally got the telescope on a star, I could then measure exactly how long they should be to bring the focus to the right spot. As it turned out, I was within 5 mm of my calculations, but the added peace of mind is well worth it. The poles ended up being 846 mm (33.3 in) in length.

I initially tried wing-nuts to tighten the blocks, but found I couldn't really get enough pressure with my fingers, so I reverted to using a normal nut and spanner. I now use a socket spanner with a screwdriver-type handle for the job. This is the only tool necessary to erect the telescope and I don't find it any problem. The nuts are never removed and so there is nothing to fiddle with in the dark – the socket slips easily over the nut and it's tight in a fraction of a second.

The Top End

I wanted to keep the diameter of the top end, the height of the focuser and the distance the focal surface is from the secondary mirror as small as possible. This is to minimise the size of the secondary mirror, and hence maximise the performance of the optics. With a low-profile focuser and minimum-diameter top end, I worked out I could use a 54 mm (2.14 in) minor-axis secondary (with only a 17% linear obstruction) and still have a reasonable unvignetted field of view, so this became my goal. (The commercial version uses a 66 mm (2.6 in) secondary with the same primary mirror.)

I decided on the classical approach of two plywood rings separated by four aluminium poles for the top end. The focuser and finder are supported on additional pieces of plywood, while the four spider vanes are supported between the four poles. Light baffling is done by lining the inside with thin black plastic sheeting. The other decision is a mechanical one – where the focuser goes in relation to the height of the top end. To properly baffle the focus, the top end has to be quite long – longer than would stow in the ball, and longer than you would really want for weight reasons. So I made the length to fit in the ball, and attached external and internal baffles to protect the focus from stray light.

To work out the minimum diameter of the top end I did the following calculations. As the mirror's focal length is 1530 mm, a 16 mm Nagler eyepiece yields 96× and a field of view of 0.85°, which translates to a linear field of 23 mm. I had already determined that this produces an excellent field and would probably be the eyepiece most commonly used with this telescope, however I decided to allow for the larger field of something like a 32 mm Erfle (48×, 1.3°, 35 mm). This means that with a 317 mm diameter mirror, you can make the inside ring diameter 317 + 35 = 352 mm without vignetting the field of view. This is one great advantage over a solid tube. A solid tube must be significantly larger in diameter than the minimum possible because of air currents along the tube which disturb the view. It is not uncommon to have a 400 mm diameter solid tube for this size of mirror

I actually made the ring inner diameter 348 mm ($13\frac{3}{4}$ in) – the resulting vignetting would never be seen and would be far less than that from the secondary mirror anyway. The annulus needs to be about 30 mm ($1\frac{1}{4}$ in) wide, giving an outer diameter of 408 mm ($16\frac{1}{4}$ in) (and therefore the minimum size for the hole in the top of the ball, although it should be made bigger than this as you have to get the ring plus focuser through – I made the hole in the ball 450 mm ($17\frac{3}{4}$ in) diameter). I laid out all the relevant dimensions on a sheet of 12 mm ($\frac{1}{2}$ in) plywood. You need to mark locations for the truss blocks on the strengthening ring and top end rings, as well as the positions for the top end poles and focuser. (You *must* mark these before you start cutting, or else you won't be able to reference positions to the circle's centre as you will have cut it out!)

I cut the rings out with a hand-held jigsaw. This is not the best way to do it, but adequate if you don't mind slightly imperfect shapes. You can smooth them later with a sander. As I've said, a router is better for this job if you have one, as it doesn't shatter the edge as much as a jigsaw.

To hold the tubes to the rings you could glue them, but this would cause problems if you ever needed to take the top end apart. To allow for this possibility, I mounted some threaded inserts into the ends of the tubes. This wasn't quite as easy as it sounds as I couldn't actually find such things – at least in my local hardware shop. The closest things I found were steel T-nuts, which had a 3/16 in threaded centre section and a spiked outer portion designed to be hammered into timber. By cutting off these spikes and carefully filing them down to the right size, I was able to force them into the ends of the tubes. You might simply glue the T-nuts into the rods of you don't want to go to the bother of trimming them to the exact size.

I was lucky enough to be able to borrow a drill the same size as the tubes which had its end ground almost flat, so I was able to drill a small depression into the rings as a guide for the tubes. This is probably not essential, but I found it very useful. A hole at the marked centre is drilled as a guide as well as for the screws. When screwing the tubes to the top end, place washers under the screws to spread the load over the wood, which is not particularly strong in compression.

The focuser and finder are attached to the top end on separate pieces of plywood (Figure 8.9, *overleaf*), held between the rings in cut-outs and then screwed to small right-angled aluminium strips. The truss blocks are screwed to the lower ring with long woodscrews. (Obviously, only one of each half is secured, the other half is free to move.)

All the wood on the top end was given a dark wood stain and then three coats of varnish to seal it. The dark

Figure 8.9 The top end.

stain shows up the pattern in the wood and looks good in the daylight and doesn't spoil your dark adaptation when observing at night. (Gloss white is the worst colour to paint the outside of a telescope tube!)

Baffling consists of lining the inside of the top end with thin black plastic. I used the covers off some computer binders, cut to shape and stuck on with some double-sided automotive tape and a few strategically placed screws. However, I made the top end to fit inside the ball during transportation and as such it isn't quite long enough to completely shield the eyepiece from direct light. To fix this problem I made a detachable extension of the same plastic which can be attached to the top end with Velcro strips.

After a little use, I decided that things appeared to be flexing a little. I wasn't sure whether it was the ball's strengthening ring or the top end, or both. The easiest to fix was the top end and so I made another ring to go between the truss blocks and the ring. Rather than make a complete ring, I made it in three sections from some scraps of plywood. These pieces were then glued and screwed to the lower ring, the joins hidden under the truss blocks. This resulted in a much stiffer ring with little increase in weight and I would recommend doing it, even though it proved that the problem was mostly with the strengthening ring.

So that the mirror collimation bolts act in the intended fashion, it is necessary that the top end always

be put on in the same orientation. This helps with minimising changes in collimation when reassembled. Some defining mark should be placed on the strengthening ring to aid in quickly positioning the top end. In my case this is accomplished with the power distribution block for power to the top end (discussed below).

Focuser

I like low-profile focusers because they help keep the secondary mirror small. I also like Crayford-style focusers because they have no backlash or play. I bought the focuser for this telescope, one of only four ready-made pieces for this project (the other three were the spider, secondary mirror and finder – although I only used the spider because I already had one). It is a JMI NGF-3. It was ordered with two options – a flat base and a short (2.5 in) focusing tube. It is expensive, both monetarily and weightwise, but worth it in my opinion. It is the only focuser I've seen that has adjustments for squaring the focuser to the optical axis to allow correct collimation.

These options are not completely essential, but make it better for this project. The flat base makes it easier to mount, while the shorter focus tube is slightly lighter, and doesn't protrude into the beam when fully wound in. This latter point is not usually a problem in normal tubes, but with the very low profile of this top end it would be.

If you use a smaller, lighter focuser then less weight will need to be added to the ball. I decided that saving about 2 kg (4.4 lb) was not worth it, but it is something to consider.

The focuser is mounted on a piece of plywood about as wide as the focuser. In it is drilled a 60 mm ($2\frac{1}{2}$ in) hole for the focusing tube to pass through, then holes for the four retaining bolts (nothing surprising here). As I've said, the NGF has four additional alignment screws to allow any construction errors to be corrected, but they need something to press against, and plywood is certainly not strong enough. I inlaid two 10 mm ($\frac{1}{2}$ in) wide strips of 1 mm thick aluminium under the focuser to give the screws something to bear against. The four retaining bolts have large washers under the nuts to also spread the pressure more evenly against the plywood.

Secondary Mirror and Support

I've already said that I settled on a 54 mm (2.14 in) minor-axis secondary mirror, but that it was only just big enough. I had to make a mirror holder that allowed full use of the mirror's surface, right to the edge (on the assumption that the mirror was good right to the edge). I haven't liked making secondary mirror holders in the past because they are small, fiddly and never seemed worth the effort, but I've also not been impressed by most of the commercial ones. The one that I did like obstructed the mirror's edge and so wouldn't work here and thus I was forced to try again at making my own. I am very happy with the design I ended up with, and it even seems to work – despite my original dislike of gluing an optical surface to a piece of metal.

The spider is part of a standard four-vane commercial one (Novak) that I had left over from an old project (Figure 8.10). I had to make new legs to match exactly the top end's diameter. These were made from

Figure 8.10 The spider and secondary mirror support.

thin galvanised steel sheet cut to size on a guillotine and then attached using the existing pins. Bolts go through the tubes that make up the top end to screw into threaded lugs at the end of the spider legs. Through the middle of the special extrusion which holds the arms is a $\frac{3}{8}$ in bolt which is used to attach the secondary mirror holder. The parts I made attach to this bolt.

When a fast mirror is used in a Newtonian, the secondary mirror must be offset slightly from the mechanical axis to catch all the light from the primary mirror and avoid asymmetric vignetting of the field. This offset is usually done by moving the spider the calculated amount within the tube – not the ideal method, but usually the only option. For this small mirror it is especially important to get it right. One of the key aspects of my design is to take this offset into account during construction so that the spider remains centred in the tube, yet the secondary is optically centred under the focuser.

The secondary holder consists of three parts; a fixed plate which is attached to the $\frac{3}{8}$ in bolt, a tilting plate which is spring-loaded from the fixed plate, and a small cylinder attached to it which holds the mirror. The fixed plate is a rectangular piece of 2 mm (0.1 in) aluminium (with rounded corners to make it as large as possible) just smaller than the mirror's cross-section. It has a central $\frac{3}{8}$ in hole where the bolt from the spider is attached and holes for three collimation screws. A matching plate of 3 mm ($\frac{1}{8}$ in) aluminium has corresponding tapped holes (I used 3 mm Allen-headed bolts). The collimation bolts are located at three of the four corners, arranged not at the more usual 120° spacing, but rather so that adjustments occur at right angles (up-down and left-right as seen through the focuser). One screw acts as a pivot and is not touched (once tensioned properly); only the other two are used when collimating. Strong springs separate the two plates to stop any backlash from occurring. On the other side of this plate is glued a short piece of aluminium tube about 25 mm (1 in) or so in diameter, the other end of which has been cut accurately at 45°. This tube is glued not at the centre of the plate, but with the necessary offset required for the particular telescope (2.8 mm or 0.11 in, in this case). Make sure to glue this to the plate with the 45° angle pointing the right way! (Guess who got this backwards the first time.)

I dislike gluing optical components, but with care it seems to work. Secondary mirrors are quite stiff because of their small size and can take a little more abuse than larger mirrors. The face of the tube where the mirror attaches was filed down so that the mirror only touched three arcs about 5 mm ($\frac{1}{4}$ in or so) long. I filed less than a millimetre away from the tube – it is only necessary to ensure that the mirror has three definite location points. Both the aluminium and the back of the mirror were cleaned with acetone to ensure there was no grease to weaken the bond, then the mirror was marked with a felt pen where the tube should attach. I did this from a template I had printed out on my computer. A ring of clear silicone sealant was applied and the tube carefully pressed into place. It was then left alone for 24 hours to cure. Once cured, I very carefully painted the edge of the mirror with matt-black paint to minimise spurious reflections.

So far, there appears to be no problem with the mirror supported in this way. Careful examination of the final image shows no sign of stress, and I am able to use all of the mirror's surface.

The Finder

Life isn't fair. The finder I selected for this telescope is both the best and worst I've ever seen. The reason I went for this finder was that it comes with a nice, quick-release dovetail bracket that allows the finder to be removed easily from the top end but still retain its alignment – a must for allowing the top end to be stowed inside the ball. The finder is a basic 8 × 50 without illuminated cross-hairs; there is a model with illumination, but I considered the option too expensive, so I fitted my own illumination system. The good points about the finder are its light weight and the bracket; the bad points were that the cross-hairs were neither centred nor focused, and the mounting bracket was far too heavy.

This last point shows how unfair I can be. Ordinarily I would praise its sturdy construction, pointing out how a finder needs to be robust and should never go out of alignment. But as I'm trying to save as much weight as I can on the top end, I found the bracket to be unnecessarily heavy. It is machined from solid aluminium, with an additional aluminium plate holding the alignment rings, and then all screwed together. A cast piece of the same dimensions with holes to lighten the whole arrangement would be preferred. By drilling a few holes in strategic places in the bracket I managed to remove some weight without compromising its strength. Not much I must admit, but every little bit counts.

Adding illumination to the finder was achieved by drilling a small hole in the side of the finder's eyepiece to allow a red LED to shine directly onto the cross-wires. (This is not ideal, but works well enough. Real illuminated eyepieces use an etched glass graticule.) Great care needs to exercised when doing this as the very fine wires are easily damaged. (Be warned – the fine wire used here I have found difficult to obtain; it is thinner than human hair.) I also repositioned the wires to make them central within the eyepiece, but I hope this was only a one-off error with the finder. A small ring was machined to hold the LED in place, and a cable was run down to a switch and variable dimmer on the top of the ball to make use of the battery already available. The cabling was broken by several plugs and sockets along the way to allow the individual pieces to be separated for storage. The only requirement is that the truss tube with the embedded wires must always be placed in the position next to the socket on the ball – hardly a problem, even when setting up in the dark.

Accessories

For some accessories it is necessary to design them into the telescope from the beginning because they affect the balance. I've already mentioned the fan built under the mirror cell and the heavy finder with its LED cross-wire illuminator that I use, but I haven't mentioned the battery that powers them. This needed to be incorporated from the start for two reasons; first, simply placing it in the ball, and second, it could be used as part of the counterweight. I decided on using only a 1.2 Ah, 12 V sealed lead-acid (SLA) battery rather than anything larger because I didn't think I'd need any other battery-powered accessories (like heating coils on the secondary mirror and finder) and so this was plenty of power. (The fan draws 110 mA when running, so there is enough charge in the battery for 10 hours' use.)

Where to put it? If it is below the centre of the ball, then it helps balance the top end, and as it is quite dense then placing it right at the bottom might seem like a good idea. But there is something else I've not mentioned – radial balance. Even more than an equatorially mounted telescope, the tube (or whatever you would call the ball, truss and top end) must be radially balanced. That means that any asymmetry around the optical axis must be accounted for. The finder, focuser and eyepiece are not on-axis, and if they are not countered by an equal and opposite weight then the telescope will not stay put. It is actually the *moment* that has to be balanced – the product of the weight and its distance off-axis, so you can balance the heavy equipment on the top end by a lighter weight (anywhere along the axis will do) but further out from the axis. It turns out that the battery can't fully balance the top end regardless of where in the ball it is placed, but it goes a long way to helping. So I made a small bracket to hold the battery near the edge of the mirror cell, opposite where the focuser and finder mount on the top end. This was put in before the ball was covered with fibreglass (I didn't have it in for the original aluminium ball – I intended to attach the battery directly to the mirror cell) by simply putting a couple of bolts through the side of the ball and covering the damage by the fibreglass. The battery is held down by two plastic straps which go through slots in the bracket.

I mounted a 12 cm (5 in) long piece of 25 mm (1 in) aluminium RHS to the top of the strengthening ring to act as the power distribution panel. I used four RCA sockets here – one from the battery, one for the fan, one for the finder, and one for a battery charger. There is a switch for the fan, and a switch/potentiometer for the finder. They are positioned for easy access from where the observer usually stands when using the telescope.

Along with the battery goes the need for a battery charger. I made one that plugs into a car lighter socket and will slowly charge the battery over a few hours. This being a portable telescope I thought that this is the best arrangement – after observing the charger is plugged in, once you awake the battery is charged ready for the next night. The charger has built-in protection and won't flatten the car's battery – it is a very useful accessory and one too often forgotten.

One addition that I haven't yet made is a light-tight shroud to cover the truss tube to help keep out stray light and body heat. It is light enough not to upset the balance, so doesn't need to be built immediately, although it is a good idea to have some idea how it will be fastened. The strengthening ring on the ball is an ideal anchor point and another reason why I made it this way.

I thought a lot about the mirror cell design and how it would be used, and decided that for maximum benefit I had to make a special tool to help with collimation. Just as my ideal is to be comfortable while observing, so I also believe that it should be possible to collimate a telescope while looking through the focuser. (In fact this ideal extends to all telescopes, not just this one.) I attached a standard socket to an aluminium pole with the aid of a small steel adapter I had made for the job. A wooden handle completes the tool. The socket slips over the threaded rod and engages the nut which moves the mirror. You can stand looking through the focuser fitted with either a special collimating eyepiece or when looking at a star and adjust the mirror without the need to move, or to need another person. It makes an easy job even easier.

A final project for this telescope is some sort of equatorial platform. While pushing a telescope around is ergonomically pleasing, I dislike having to keep doing it in order to keep an object centred. This is why most of my telescopes are on equatorial mountings. This lack of tracking ability hinders high-powers and looking at planets and is my only dislike of this telescope. I have plans for a platform which does away with the need for grinding surfaces by using rolled tubing to provide the running surfaces.

In Use

To assemble the telescope from being packed in a car into working condition takes just 2 minutes. The sequence is to first place the stand where you want to observe, then carry the ball (complete with top end) onto the stand. The strengthening ring makes a perfect handle. The top end is removed from the ball and set aside. The six truss poles are removed from their carry bag and inserted into the holes on the ring, making sure that the pole with the power plug goes next to the socket. Next the top end is placed atop the poles – it is easier to manoeuvre this into place before tightening the lower truss clamps. Check that the focuser goes in

a b

Figure 8.11 Setting up the telescope.

the right position, too. The top end on, the six nuts are tightened to lock the truss in place. The finder is removed from its protective box, placed on its mount on the top end and plugged in. Once the baffle extension is placed on its Velcro strips, the mirror cover is removed and it is ready to use.

The telescope has lived up to my every expectation and is a delight to use. While it took me longer to make than any other telescope I have made (even excluding the time to make the mirror), it has been a very satisfying project.

c

d

e

f

Part III

Mounts

Chapter 9
Equatorial Platforms

Chuck Shaw

Since the late 1970s, equatorial platforms have permitted Dobsonian telescopes to track equatorially. Equatorial platforms tend to keep to the same ATM philosophy as that of the Dobsonian mount: simple and buildable with basic workshop tools. Chuck Shaw has improved upon the original platform of Adrien Poncet to enable accurate tracking and guiding for photography and CCD imaging.

The popularity of the Dobsonian type of mount has caused a virtual revolution in amateur telescope making. The only thing it lacks is the ability to track the stars automatically. At low magnifications this is not a terrible burden. Gentle nudges easily keep the target in the field of view. However, at higher magnifications it becomes more challenging. Again, this is not unduly difficult when you are observing by yourself, but it is at its worst when sharing the view with others or when trying to make eyepiece impressions (sketches). A simple way to have a Dobsonian track the sky would be very convenient.

This motivation prompted the invention of the equatorial platform. Simply stated, it is a platform sitting on bearings that rotates at sidereal rate (or solar or lunar rates) (Figure 9.1, *overleaf*). Sit your Dobsonian on top of it and have the best of both worlds: the ease of construction and use of a Dob and an easy to build and use tracking system. Even though the intent and tracking accuracy of the platform is designed primarily for visual observing, simple

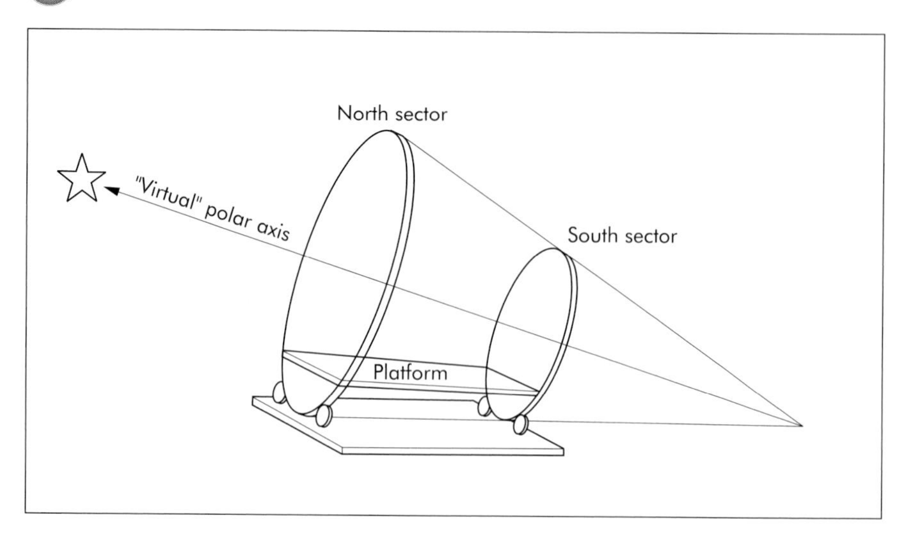

Figure 9.1 The principle of the equatorial platform.

astrophotography is possible if the platform is carefully constructed and polar-aligned.

An early design for an equatorial platform was documented by Adrien Poncet (Poncet, 1977). It uses a pivot point and rollers/slides on a plane to define the motion of the platform. Alan Gee designed a platform that used a cylindrical bearing on one end and a single pivot on the other. This was an improvement on the high loading of the Poncet design, but the "virtual" polar axis was very low and the system was not perfectly balanced, which required more power to turn. Georges d'Autume changed these designs by eliminating the single pivot and introduced the concept of using conical bearings to carry more weight and also raise the polar axis higher for better balance. Georges d'Autume provided an excellent review of all these designs back in 1988 (D'Autume, 1988). Platforms based on this conical bearing design work well, but it is not straightforward to fabricate conical bearings! There had to be a better way!

One of the advantages of building a Dob is that it only requires average building skills and only basic, hand-held power tools. The platform, to be compatible with the idea of a Dob, must keep to that same philosophy. That is what prompted changing the conical bearing design to use cylindrical bearing surfaces for the north and south bearing surfaces, similar to the north end of the Gee design. Before we go further, let us review how a platform works.

How Does It Work?

One way to visualise how an equatorial platform moves is to imagine a cone lying on its side with its centre axis (polar axis) aimed at the north celestial pole. The side of the cone is parallel with the ground (the $\frac{1}{2}$ angle of the cone will be the same as your latitude). Put bearings under the cone so it will roll around its axis. Inside the cone, attach a flat platform just above the bearings, as shown in Figure 9.1 . Now in your mind's eye erase the entire cone except the part under the platform that rides on the bearings, and the platform itself. This is the shape of the conical bearing surfaces under the platform that Georges d'Autume suggested. The axis of rotation is the centre-line of the old cone (before you erased it). Polar alignment of the platform aims this "virtual" polar axis at the north celestial pole (the same as the earth's spin axis). Rotate the platform in the opposite direction to that in which the earth is rotating, and the platform stands still (in rotation) with respect to the stars, as does anything that is sitting on the platform (like your telescope). The design works the same for both northern and southern hemisphere observers by simply reversing the rotation of the platform, and switching the references to northern and southern portions of the platform in the descriptions.

If you "shorten" the cone by placing its apex at one edge of the platform, and its base at the other edge of the platform, and replace the base of the cone with a disc that is the same diameter as the end of the cone, you get the modification that Alan Gee made to the design. It makes the northern bearing surface a cylindrical surface, which is much easier to make, and uses the single-point swivel at the apex from the earlier Poncet design. This design is very robust and easy to build, but has one potential drawback. The imaginary line of the virtual polar axis is very low as it passes through the scope and rocker box, well below the centre of mass of the scope and rocker box. This does not pose problems for a drive system mechanically coupled to the platform. However, it does require a larger motor than if the axis passes through the centre of mass of all the components that are rotating. If you want to drive the system by attaching a motor to one of the bearings (to use it as a drive roller), slippage can become a problem if the system is not balanced about the axis of rotation. An unbalanced system requires more force to rotate, and the friction of the roller against the sector must exceed this required torque or the system will slip and not track properly.

To raise the rotational axis high enough to go through the centre of mass of the rotating parts, go back to the longer cone that the d'Autume design uses, which allows the polar axis to be much higher. Using the same type of approach as the Gee design to get cylindrical bearing surfaces, insert two discs inside the cone that touch the northern and southern ends of the platform inside the cone, as shown in Figure 9.1. These two discs (or very short cylinders) have their centres on the virtual polar axis and extend below the platform inside the cone. Again, erase everything above the platform and what is left is an all-cylindrical bearing equatorial platform. The conical bearings are entirely gone (as with the Gee design), and the polar axis is up where it needs to be, through the centre of mass of the rotating system (as with the d'Autume design). You must fabricate a second cylindrical sector, but this is actually very little extra work. The benefits of the balanced rotating system versus the slight increase in construction work building a second sector are an excellent trade-off to consider.

All platforms rotate the scope and mount that sit on them. If rotated too much, the scope will fall over. However, if you limit how much rotation is allowed, it never becomes a problem. Most designs limit the rotation to approximately 7.5° on either side of vertical. This allows a total of 15°, which provides approximately 1 hour of tracking.

Making It Track

There are a number of ways to drive the platform. The simplest concept (and actually the most accurate) is just to turn one or more of the bearings into a drive roller that is connected to a motor. While simple in concept, this option is unfortunately a bit more complicated to build. You have to use materials for the roller to avoid/minimize slippage and wear (very hard rollers, of for example steel or stainless steel, and softer sectors, of perhaps aluminium or aluminium-faced wood, are a good combination). You also need a clutch to disengage the drive motor to roll the platform back to its beginning of travel or a way to lift the sector off the drive rollers to move it back to the starting point. Just skidding the sector across the roller will work, but you run the risk of damaging the roller surface. Balance is also more critical, since friction is the only thing keeping the sector from slipping on the roller. However, a system like this can work really well and consistently allow 15 to 30 s unguided images with a CCD camera at prime focus, and much longer exposures if guided (in RA).

Slightly less sophisticated, but easier to build and very forgiving of errors in balance, is a tangent arm drive. This option has a drive screw that has a carriage (nut) on it, which grabs a tangent arm attached to the platform. The linear motion of the carriage along the drive screw is turned into rotational motion. The carriage has a tang attached to it that grabs the tangent arm. The tang has a vertical slot to allow for the motion of the tangent arm. This design is only perfectly accurate near the mid-travel point of the platform, when the drive screw carriage is actually tangent to the arc the tangent arm describes. However, the tracking inaccuracies due to this "tangent error" are actually very small, owing to the large diameter of the sectors involved. All these extra pieces each introduce a bit of play into the drive system, which can add up and hurt tracking accuracy. In addition, imperfections in the drive screw and carriage can cause periodic errors. However, for visual work, including using high power, the accuracy is more than adequate, and you may find you simply do not need anything more sophisticated. For piggyback photography this approach is also just fine, since these tiny errors do not show up in the larger scale images. For unguided prime-focus photography, the combination of mechanical play of all the pieces and periodic error from the drive screw usually limits exposures to about 15 s (which is still OK for CCD imaging where many short exposures can be stacked). For any long-duration photography, with a platform or any other type mount, you really need to guide the tracking (either manually or automatically) to obtain consistent quality in exposures.

Another way to attach a drive screw to the tangent arm is by a wire or chain bent around a "sector" instead of just a single tangent arm pin. This provides an increase in accuracy at very little additional complexity, even though periodic errors associated with the drive screw are still present. The biggest problem is that you cannot lift the upper platform off the lower platform because they are connected by the wire or chain. Both of these options also require a clutch between the motor and the drive screw. To provide a simple, cheap clutch option, use plastic gears like those in radio-controlled cars, and pivot the motor to disengage them.

Yet another drive design to consider is to have the north sector ride directly on top of the threaded rod. The sector needs to have threads cut into its edge. Apply a thick bead of epoxy putty along the outside radius of the sector. Just as the epoxy is starting to get firm, take the threaded rod and, with a rocking motion of the sector along the threaded rod, carefully impress threads into the putty. You can also press the putty directly onto the straight rod, and just before it hardens remove it from the threaded rod and wrap it around the sector. I do not recommend bending the threaded rod to impress the threads into the bead of epoxy already in place on the sector, owing to the difficulty of achieving a very accurate bend in the threaded rod. Even tiny errors will show up in the eyepiece or camera. Also, coat the threaded rod with some anti-stick cooking pan spray to prevent the putty from sticking to it.

The bottom line is that there are a lot of ways to impart the required motion to the platform! These have been but a few. I suggest first building the very simple tangent arm drive. It is simple and more than accurate enough for even very high-power visual work when the platform is polar-aligned and running at the right speed. Then, work to minimize the mechanical play in the system and to remove or reduce the periodic errors from the drive screw to be able to do short-exposure photography. Later, after enjoying using the platform for a while, if you want to try to further improve the tracking accuracy, consider a drive roller system that replaces one or two of the bearings. The beautiful secret to this approach is that you initially build simple things, and can be using and enjoying the platform as you build upgrades to it!

There are also a number of options on how to power the platform. Battery power frees you from needing household a.c. power. However, if you do not mind being tied to a power source, then a.c. synchronous motors will solve the problem easily (more below on how to decide what speed). A.c. synchronous motors also have the advantage of being able to use easily available (or home-built) drive-correctors to vary the frequency (and thus the speed) of the motor for manual or automatic guiding. If you favour battery operations, then you must decide on either a d.c. stepper motor or a regular d.c. motor. A stepper motor is very accurate. It only rotates a measured amount each time its windings are energized. Then the next set of windings are energized and it moves again, and so on. Reasonably simple circuits like the one in Figure 9.9 (p. 149) are easily built to drive stepper motors. However, remember the steps have to be fairly small and fast to not be seen when using the scope visually. This constraint, in turn, requires gearing down the stepper motor to allow it to turn faster. Use stepper motors with 1.8° per step (200 steps per turn) if possible. I suggest running the motor at least 40 steps (or half-steps) per second or faster to keep the vibrations from the steps from being objectionable. Circuit designs that "half-step" (make the size of the steps smaller) also really help smoothness and precision. The stepper motor controller circuit in Figure 9.9 will half-step the motor as well as reverse the direction of the rotation (handy for having the motor rewind the drive screw when it is at the end of its travel).

There is another very simple option that sacrifices only a small bit of accuracy, but is still more than adequate for high power visual work. Simply run the platform off of a regular d.c. motor. Vary the voltage to change the speed of the motor. As the battery wears down, increase the voltage with a simple potentiometer (pot) in series with the motor to speed it up. Figure 9.8 (p. 148) shows a very simple automatic voltage regulation circuit based on a Radio Shack (Tandy) variable voltage regulator that you can use to automatically maintain the d.c. output voltage constant. This is actually accurate enough for some photography! Consistent with the "simple first, then upgrade" philosophy for the mechanical drive considerations, take the same approach for the choice of motor drive. Due to the simplicity and lower cost of using a d.c. motor and a simple manual pot to control motor speed, start with this option. Later, experiment with adding the automatic voltage regulation circuit. This way you also have a very durable and simple backup system for powering the d.c. motor. Then upgrade later to a stepper motor if you feel you need it or just want to build it. Again, the idea is to start simple, and then add upgrades while you enjoy your platform.

All this theory is good, but how do you bring this to life? What follows will describe the different steps to build a platform. Read the rest of this chapter a couple of times to better visualise the actual activities. For those of you who have read this far, but have concerns about your abilities to make something like a tracking platform, I assure you that after you build your first platform, using only simple hand tools, and watch it magically track the stars with high accuracy, you will look back and wonder why you thought it was such a big deal before building it, and how you ever did without it! So, enough with just thinking – let's just go do it!

Design and Construction Tips

The first step is to measure the height of the centre of mass of your scope and rocker box. If you intend to just sit the entire scope and Dob mount on top of the platform, then measure the height from the centre of the altitude bearings all the way to the ground. Add about 2 in (5 cm) to allow for the sectors under the platform. If you intend to use the moving platform itself to replace the old Dob ground board (to lower the overall height of the scope slightly), the dimension is from the centre of the altitude bearings to the top of the current ground board (again, add something for the sectors under the platform). Also measure the size of the ground board (length and width) to determine the spacing between the cylindrical bearing sectors. You also need to decide the latitude where the platform will be primarily used. North or south of that latitude it will still work, but will need to be shimmed to keep the virtual polar axis aligned with the north celestial pole. Five degrees of shimming will cause no problems. This equates to about 300 miles (480 km) north or south. Up to 10° will work, but the scope stays tilted quite a bit. Round off the latitude to the nearest degree to make measuring and cutting the pieces easier. Using a powered table saw for cutting the pieces will help keep things accurate. However, the design is actually pretty forgiving for most of the design and building activities, especially if you are not trying for photographic tracking accuracies, so reasonable care in measuring and cutting should be sufficient.

Lay out the platform design in a scale drawing similar to Figure 9.2. Be reasonably careful, but again, perfectly exact dimensions are not really required at this stage. For your drawing, assume the base board is $\frac{3}{4}$ in (18 mm) plywood, and the sectors extend 2 to 2.5 ins (5 to 6.5 cm) below the bottom of the plywood. Now measure the radii of both the north and south sectors from your drawing. If you are better at math than at scale drawings, you can use the three diagrams at the end of the chapter (Figures 9.10, 9.11 and 9.12) to calculate the sizes of the platform and sectors. They were developed by Ed Grafton when he laid out the plans for his platform for his 18 in (457 mm) Dob. Otherwise, you can just use the dimensions he came up with for his platform if your scope is about the same size or smaller. It is important to note that the centre of mass of the scope and rocker box does not have to be between the two virtual cylinders. At very low latitudes, where the sectors are more vertical, this will almost always be the case unless the rocker box is

Figure 9.2 Cylindrical bearing equatorial platform (view is from the west, looking towards the east).

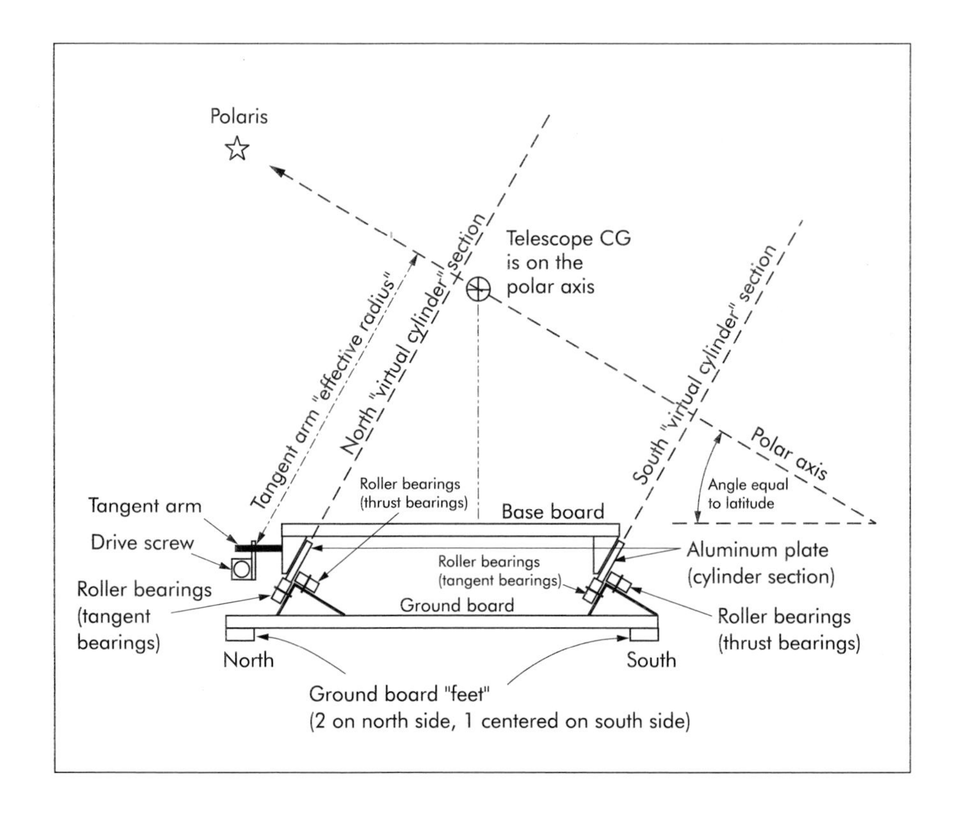

very tall. At higher latitudes, as the sectors lay back more and the virtual polar axis is aimed higher, the centre of mass may be out in front (to the north) of the plane that the northern sector describes. The goal is only to have the centre of mass of the rotating hardware on or very near the virtual polar axis.

Making the sectors from metal is a bit more work than just cutting them out of wood with a router and then covering the edges with thin metal or Formica. However, I think the increased durability is worth the extra effort if a rather heavy scope will be sitting on the platform. For very lightweight scopes, all wood sectors cut out with a router will work perfectly. Use the hard rubber wheels from in-line skates or skateboards if the sector is wooden and the sector surface will not be faced with metal or Formica. If the scope is slightly heavier, but still not a heavyweight, say less than 50 lb (22 kg) then the wooden sectors really need to use bearings that are harder than the rubber wheels (the rubber wheels will distort slightly under the added weight and that will detract from the tracking). Perfect bearings are available from in-line skates or skateboards, just pop the bearings out of the hard rubber wheels, or better still, just get a new replacement set of bearings (most come in sets of eight bearings, which is how many you need anyway!). These bearings are metal and can cause small dents in the wooden sectors. For this reason, surface the sectors with either smooth Formica for lighter-weight scopes, or, better still, use thin aluminium ($\frac{1}{16}$ in thick by $\frac{3}{4}$ in) (2 mm × 19 mm). Attach the Formica or the aluminium with spray-on contact cement (the spray-on type insures a thin, even coat). Use small screws at the very ends of the facing material to secure the facings better than with just the contact cement. Take care to drill pilot holes for the securing screws, since you will be screwing into the end grain of the wood and it can split quite easily. Also, face the side of the wooden sectors that the thrust bearings ride against. A metal bearing rolling against a wooden surface will create small motions that will be seen in the eyepiece at higher powers.

Some builders have used pebbly Formica like Ebony Star for the sides of the sectors and Teflon blocks instead of bearings for the thrust bearings. This will add friction, and should not be used for drive roller systems which need to rotate with very little friction. However, it should work well for very lightweight scopes rotated with a tangent arm drive.

The sectors themselves, whether wood or aluminium, should be attached to two blocks of triangular cross-section under the moving upper section of the platform. The cross-section shape will need to match the latitude the platform is built for. For example, at 40° north latitude, the sectors will be canted back 40° from vertical. Scribe the arc for the sectors on the blank material. For wooden sectors, a router with a jig to cut the arc works well. Use a hardwood like maple if you make the sectors out of wood. For metal, scribe the arc and cut it out with a hacksaw in a series of small cuts (or much better, a bandsaw, if you can get access to one). Take care to not cut "inside" the arc. For metal sectors you will need to set up a grinding jig to sand the sectors smooth.

Attach the sectors to the support blocks, and then attach the support blocks to the base board. Since there will be bearings riding against the back sides of the sectors, do not place any screws holding the sectors to the support blocks within about 1 in (25 mm) of the outside curved edge of the sectors. It is *very* important to make sure the side surfaces of the north and south sectors are perfectly parallel. This is to insure that all the bearings remain in contact throughout the swing of the platform. To assure this, permanently fasten one support block/sector to the platform, but the other support block should not be glued, or otherwise permanently fastened. Just screw it down as accurately as possible for now and carefully mark where it is located. Later, when adjusting the mating of the upper and lower portions of the platform, loosen these screws and make tiny adjustments as needed.

If the sectors are metal, they must be smoothed from the coarse hacksaw/bandsaw work. In order to do this, you need to build a jig to hold the base board as shown in Figure 9.3 (*overleaf*). Use something like 1.5 in (38 mm) electrical conduit or a heavy iron water pipe as the axle for the jig to rotate on (it will lie along the virtual polar axis). Make the $\frac{3}{4}$ in (18 mm) thick plywood jig of a size to hold the conduit/pipe the right distance from the base board so the conduit/pipe is exactly along where the virtual polar axis is located, which is at the centre of the circle that the sectors are a portion of. Make sure the *centre* of the conduit/pipe is along the polar axis, *not* the *edge* of the conduit. To do this, the jig will have to be offset to the side from centre of the platform to make the *centre* of the conduit ride along the centre of the virtual polar axis. This is

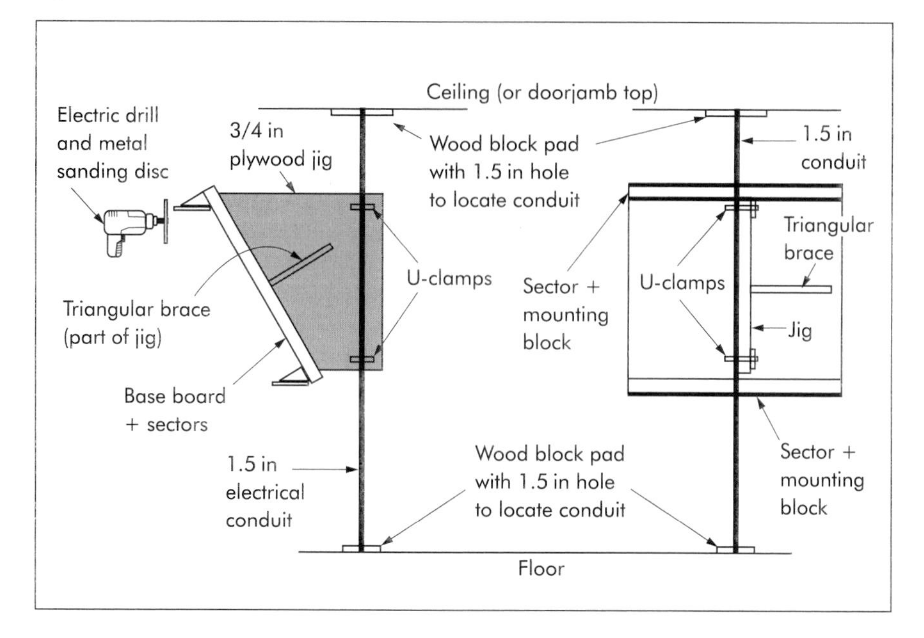

Figure 9.3 Jig for smoothing the platform sectors.

important, and really easy to forget to do! Attach the conduit/pipe to the jig with two U-bolts. Cut the conduit/pipe to a length to allow it to be tightly wedged into a door jamb or garage door opening. Keep it from wandering around with two wood blocks with 1.5 in (~38 mm) holes in them tacked to the door jam and floor. The jig needs to keep the base board square, so add at least one triangular brace between the jig and the base board. All this sounds complicated, but it's really simple. If more than one platform is being made, it is also shared work, since the jig is reusable. If the U-bolts do not hold the conduit in place well enough (that is, with no play), add wood blocks snugly beside the conduit to better locate it on the jig. Angle iron or angle aluminium is even better than wood blocks. Add a clamp on the conduit/pipe below the jig to keep the jig at the right height (use a *big* fender [mudguard] washer between the jig and the clamp).

The metal sanding should be done with a portable electric drill and a metal sanding disc, or a belt sander with a heavy-duty sanding belt. The electric drill or sander should be held in place securely, as shown in Figure 9.4. Use something like a Workmate or a sturdy sawhorse or a convenient table or table leg to clamp the drill or sander to. It must not be allowed to move while sanding! Rotate the baseboard/jig back and forth while just lightly touching the sanding disc/belt. Do not try to

Figure 9.4 Smoothing the platform sectors.

take too big a bite with the metal sanding disc, since it will deflect and you will end up with a surface that is not square. If this does happen, the platform will still work just fine, the bearings will just not ride on as much surface of the sector. Go very slowly, taking small portions of the sector off at a time and enjoy watching the fine finish appear. You will be amazed at the precision grinding that will be the result from using such a crude setup! Do not stop till all the saw and sanding marks are totally gone from the sector surface and the contact with the sanding disc is uniform all the way through the swing of the sector as it passes the sanding surface. Then do the same thing with the other sector without moving anything on the jig (except perhaps the height of the clamp on the conduit/pipe to lower the baseboard/jig to line up with the clamped sanding machine). Also take the time to observe the motion of the platform around the polar axis (conduit/pipe). This is the same motion the platform will describe while tracking. Make sure you are wearing glasses or protective goggles while grinding!

With the upper portion of the platform completed, start work on the ground board. The bearings can be mounted using wooden blocks or 2 in (50 mm) aluminium angle. You have to whittle on the shape to get the tangent bearings (that ride against the outside edges of the sectors) to be in the same plane as the sectors, and then do a bit more whittling to get the axles of the thrust bearings (that ride against the sides of the sectors and keep the edges of the sectors in place on the tangent bearings) to cant inward towards the virtual polar axis. Unless you do this, the rollers will skid as the sectors roll back and forth. Go slowly and compare often and the job is easy. If you are using wooden sectors, try using the concave scrap leftover as a mount for the rollers. It already has the approximate shape you will need to align the bearings; just cut the bottom off at the same angle as your latitude and use angled wooden blocks to further brace it in place. You can also cut a concave wooden piece that matches the metal convex sectors and mount the aluminium angle on it to get the roller orientation right. The key to mounting the bearings is to aim the axles for the thrust bearings at the virtual polar axis, and to have the tangent bearings in the same plane as the sectors (that is, so they roll evenly on the outside edge of the sector).

When all four bearing holders are ready to be mounted, place the north bearings in place (they will be wider apart, since that sector has a slightly larger diameter). Then place the south bearings in place. Move the bearings in and out till you have the same amount of rotation in each direction. You will have to play with the exact location of the bearing assemblies to make sure that all eight bearings stay in contact with their respective sectors at all times. The secret is to go slow and be patient. Remember, the four bearings that ride against the south sides of the two sectors *must* have their axles aimed inward at the polar axis, or they will not roll correctly! The four bearings that roll against the edges of the two sectors have their axles parallel with the polar axis.

If you find you have played with getting all the bearings to remain in contact throughout the entire rolling travel but simply cannot get that to happen, it may be that the two planes that the sectors describe are not perfectly parallel or that the centres of the cylinders for each of the sectors are not both on the same virtual polar axis. Try loosening the mounting screws for the sector mounting block that was not permanently attached and adjusting its position. If the bearing that does not remain in contact is a tangent bearing, keep playing with it till all of a sudden everything remains in contact (rest assured: if your sectors are reasonably round then there *is* a place where this will happen!). If one of the thrust bearings loses contact and you cannot get it to stay in contact no matter what, move both thrust bearings away from the south sector (or move the south sector closer to the north sector) a very small amount ($\frac{1}{16}$ in or so) (2 mm). Some builders have even reported success with omitting the two thrust bearings that ride against the south side of the south sector. This will, in effect allow the plane of rotation to be totally defined by the plane of the north sector. The reason I do not suggest this as the normal default configuration is that if the scope/mount gets kicked or bumped then the extra two thrust bearings on the south sector (even if not in contact, but riding very close to contact) help keep the upper moving platform/sectors from getting cocked and falling off the tangent bearings. If you take this approach, you could get this extra insurance with Teflon blocks for the southern thrust bearings instead of using bearings.

The ground board should have three feet, to ensure that the platform will not rock. Two are on the north side, the third centred on the south side. When setting the platform up and polar aligning, it will be important to level the platform in north/south tilt – more about this below. Building adjustable feet is a good idea to make tilt adjustments easier. Use T-nuts on the underside of the bottom board and $\frac{3}{8}$ in (M10, 10 mm) carriage bolts. Put a 3 in diameter (75 mm) wooden disc on the end of the carriage bolt as a foot so that it won't sink into the ground, and run a nut and washer up against it. Add a small bull's-eye bubble level that is adjustable with shims. After the first really good polar alignment has been done, you can adjust the shims so that the level indicates level. For subsequent coarse polar alignments, you can quickly adjust the feet to get the polar axis tilt very close and only really have to worry about azimuth alignment!

The tangent arm is next. A $\frac{1}{4}$ in (6 mm) diameter or larger lag bolt (carriage screw) with its head cut off will do nicely (screw it in first, *then* cut off the head!). Either measure from your drawing or use the grinding jig to measure the radius the tangent arm describes. Then calculate the circumference of the circle the tangent arm makes (remember: $2 \times \pi \times r$ = circumference). Divide the circumference of the tangent arm's circle by 1436 to get the units per minute that the tangent arm must travel (in the same units the circumference was stated). I use a $\frac{3}{8}$ in diameter, 16 threads per inch all-thread rod ($\frac{3}{8}$ in UNC studding) as a drive screw. That means the rod must turn 16 times to move the tangent arm 1 in. Multiply the required speed of the tangent arm in inches per minute, by the number of threads per inch, to get how fast the threaded rod must rotate per minute. If your circumference was in mm and the threaded rod is in threads per inch, divide the threads per inch by 25.4 to get threads per millimetre. For average-sized platforms the rotation speed of the drive screw will usually be somewhere between 1.5 and 2 revolutions per minute.

Mount the threaded rod between two bearings, as shown in Figure 9.5. Attach each bearing to the rod between two nuts jammed against the bearing. Make sure the nuts bear only against the inner races of the bearings and not the outer races. A nice drive box is made by a wooden rectangle $\frac{3}{4}$ in (19 mm) thick and about 3 in (75 mm) wide and 12 in (300 mm) long with two end pieces attached (like bookends) the same width and about 4 in (100 mm) tall, with holes drilled in the ends the same size as the bearing outer races. Pieces of 2 inch (50 mm) aluminium angle can also be

Figure 9.5 The tangent arm drive.

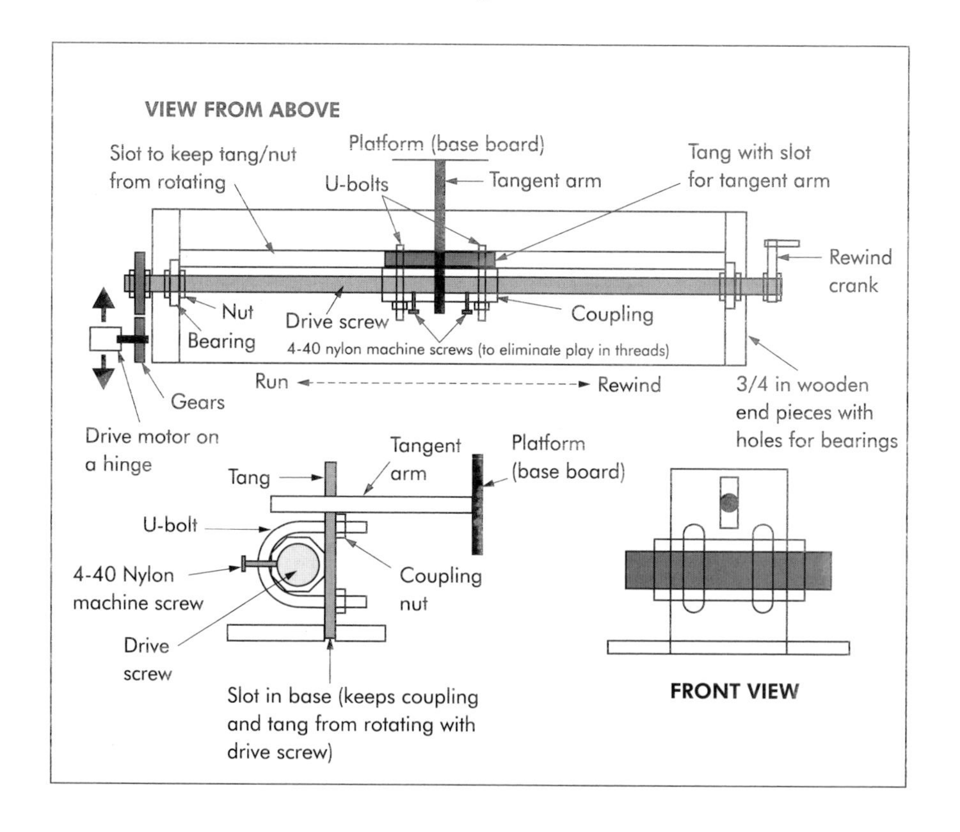

used to hold the end pieces in place. Holes in these end pieces that are same size as the outer bearing races hold the bearings. Use large fender washers on the inside of the drive box ends to keep the threaded rod and bearings in place in the holes and to resist the pushing and pulling on the platform that the drive screw must perform. The centre holes of the fender washers need to be large enough for the nuts that clamp against the inner races to fit inside the holes. Attach the fender washers to the insides of the box with small bolts or screws. You will have to adjust exactly where along the threaded rod the nuts clamp the bearings in order to not create too much friction in the bearings, yet not have any end play in the threaded rod. The length of the travel of the drive screw must be adequate to swing the platform through 15° (that is, 1 hour of travel). Use gears found at hobby shops for electric cars to couple the drive screw to the motor. Two nuts and flat washers on either side of the gear on the drive screw will hold it in place without having to have a hub. Make sure it has no wobble after it is tightened (the centre of the gear needs to be at the centre axis of the rod). Mount the motor on a hinge to be able to swing it into mesh with the gear on the drive screw, and to swing it out of the way to disengage it during rewind. Hold it in mesh with a small spring.

The nut that travels along the drive screw should be long, like a coupling for the all thread, to provide stability. Attach a flat plate (tang) to the coupling with two U-bolts. The plate needs to be of relatively thin material, such as $\frac{1}{8}$ in (3 mm) aluminium. Make the plate long enough to reach down into a slot formed by two runners. The slot will keep the tang (and coupling) from turning with the drive screw. Cut a slot in the plate for the tangent arm to fit through. Don't make the slot too big, or the play will show up in the eyepiece! However, the thicker the plate material, the looser the slot must be, since the tangent arm will be at a slight angle at the ends of the travel and will bind. A strong spring attached between the tangent arm and the plate to always hold the tangent arm against one side of the slot will eliminate the play. You need a slot instead of just a hole, since the tangent arm describes an arc that has it lower at mid-travel than at the ends of travel. Ball joints or even socket wrench universal joints can be used very effectively to couple the tangent arm to the tang. Copy machine repair shops are a good source of small gears and bearings and things like the ball joints.

Be sure to allow for the vertical motion to happen if not using a slot.

If the coupling is too loose on the threaded rod (too much play in the threads), there are at least two solutions. One is to pack the threads in the coupling with a mixture of talcum powder and epoxy, and coat the drive screw with cooking pan no-stick spray, or silicone lubricating spray. Then, slowly screw the drive screw into the coupling and let the epoxy harden. The talc/epoxy mixture will have made snug-fitting threads for you and the lubricant will not allow the epoxy to bond to the drive screw. An alternative is to drill and tap two very small (no. 4 machine screw, 40 threads per inch or a similar small size) holes into the coupling at each end. Insert nylon machine screws into the holes and gently tighten them to eliminate any play. Both methods work wonders in eliminating play which will find its way into the eyepiece view. Its also a good idea to first coat the threaded rod with fine-grade valve-grinding compound (auto parts shops carry it) and "lap" the nut and drive rod by screwing the nut back and forth several times along the drive rod. Amateur telescope makers with mirror-grinding compounds can make their own lapping compound from a mixture of Vaseline (petroleum jelly) and something like no. 320 grit grinding compound. This will eliminate any tiny burrs that will make tracking rough and show up in the eyepiece. The drive screw box and motor can be mounted on the ground board. Make sure the box is not too close to the platform, or else the platform corners will hit it at the end of the travel. Figures 9.6 and 9.7 show the finished platform and ground-board.

You can use the upper moving board as the ground board of your Dob mount, or you can simply set the whole mount on top of the platform. If you use the platform for the ground board for the Dob mount, you may want to put a couple of long, $\frac{1}{4} \times 20$ ($\frac{1}{4}$ in UNC) bolts and T-nuts between the base board and ground board on the east and west sides to hold them together when you transport the whole thing. Don't forget to remove them before trying to run the platform though or it will stall!

The motor controller can be a simple battery and a potentiometer hooked in series with a d.c. motor. In fact, I carry a spare emergency controller in my parts box made of these components and have loaned it to friends who have had problems in the field. Start with

Figure 9.6 The platform and the ground board assembly.

Figure 9.7 The platform on the ground board assembly.

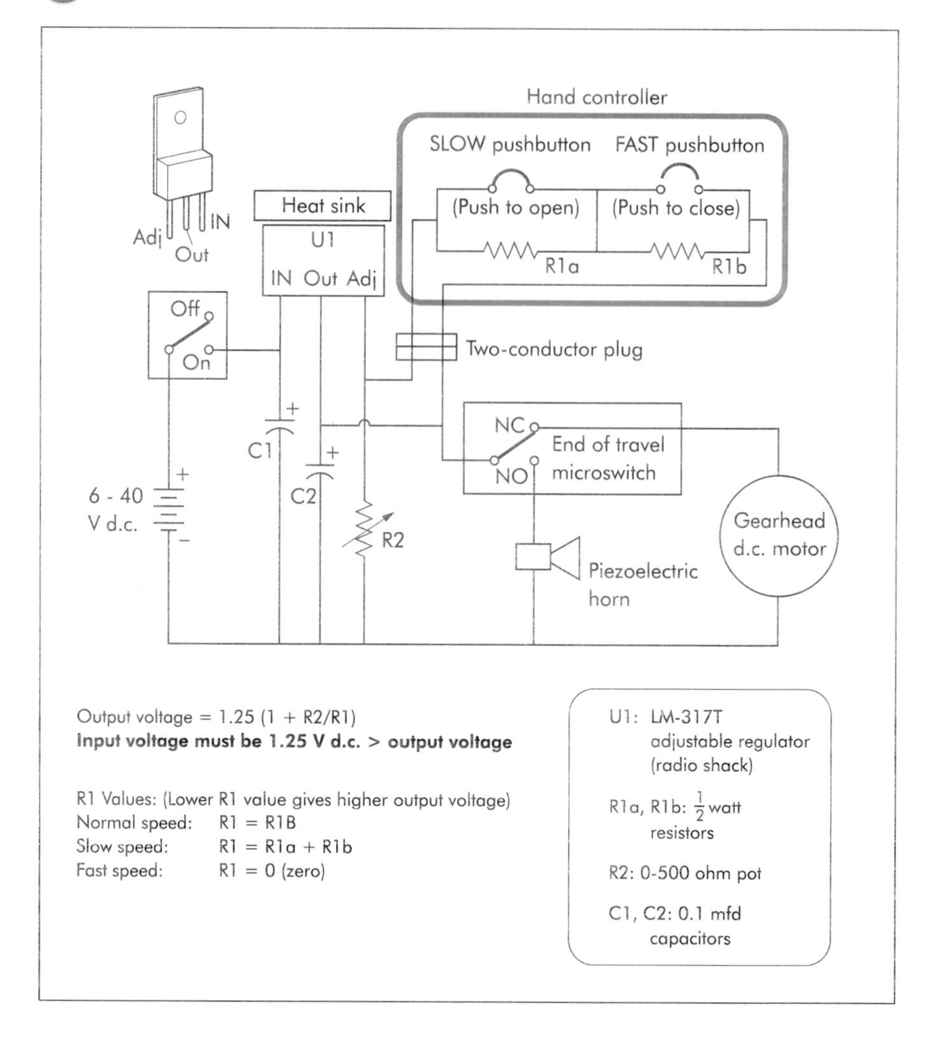

Figure 9.8 Automatic voltage regulator.

something like that to get running as soon as possible. When you are ready, build an automatic d.c. voltage regulator like shown in Figure 9.8. All parts can be purchased at an electronics store like Radio Shack (Tandy). Hold all the parts in place with a small piece of perforated circuit board, and put it all into one of the small experiment boxes (also found at Radio Shack/Tandy). A d.c. motor allows you a great deal of latitude in gearing options, since you can vary the speed through a large range. If you use a stepper motor you can use a simple stepper controller circuit as shown in Figure 9.9. You will need to make sure the stepper motor rotates fast enough to have at least 40 to 60 steps per second to avoid seeing the steps in the eye-

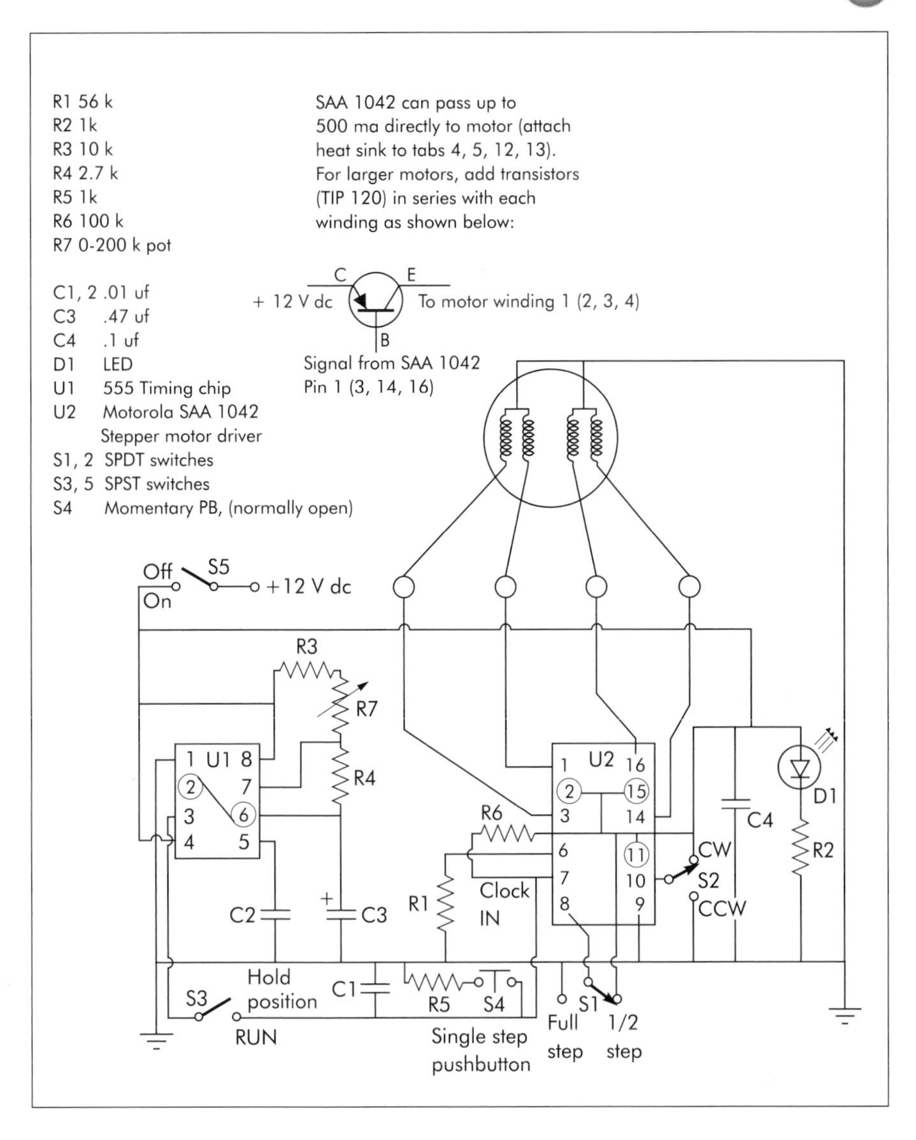

Figure 9.9 Stepper motor controller.

piece. That means, for a 200 step (1.8°) motor that is being half-stepped (so it makes 400 half-steps per turn) the motor will need to be turning about 6 rev/min. Since the drive screw needs to be turning about 2 rev/min , that means a gear reduction of at least 3 or 4 to 1. If you use an a.c. synchronous motor the gear ratios need to be fairly closely matched to what the drive screw needs to turn at since, even with a variable-frequency drive-corrector (such as the one described in Chapter 15) the speed changes available to tune the speed for the motor are limited.

To use the platform is simplicity itself. Align the virtual polar axis with the earth's polar axis and turn it on. Adjust the motor speed to eliminate any right ascension rate error and enjoy! Once you get the platform initially polar aligned, attach the level to the ground board (the south side is best I think) and shim it till it reads level. It is important to note that when properly aligned, the upper platform board will be level only at the centre position. The bubble level allows repeating the north–south tilt setting of the platform very quickly. To repeat the azimuth orientation of the platform, use a small compass. Attach the compass to one end of a 12 in (30 cm) by 3 in (7.5 cm) board. Hold the board up against the ground board side and read the compass. The board is to hold the compass away from the metal of the platform so that it reads correctly. Remember, though, if you move to a different location, the compass and level may not be accurate for that location! I also use the finder on my scope. I have pins on the altitude and azimuth bearings that lock the scope's optical axis with the platform's virtual polar axis. Then I just move the whole platform and telescope around till the pole is in the right place in the finder's cross-hairs. (This is only really accurate if the platform has replaced the Dob's ground board.)

The best way to achieve a really accurate polar alignment, whether for an equatorial platform or any other type of tracking mount (such as a German equatorial mount) is the two star drift alignment technique (a good description is on the *Sky & Telescope* web site). When you have the platform accurately polar aligned, remember to attach the level and the compass. Any errors from then on are due to the motor running too fast or too slow. The d.c. voltage regulator will automatically adjust the voltage for you as the battery slowly drains (at least till the battery gets to about 1.5 V d.c. above what the motor needs, after which it cannot help any more). Most d.c. motors I have used draw about 30–50 mA and are 12 V motors that I am running at about 3 V d.c. I use rechargeable 6 V d.c. gel cells from a surplus store, or a normal lantern battery. A lantern battery lasted an entire week once at the Texas Star Party!

If you want to do unguided photography, or guided in RA only, the motion is fine and has no field rotation problems anywhere in the sky. To guide in both RA and Dec, it is best to stay within 10° to 15° of the meridian to avoid field rotation due to declination corrections. Make declination corrections when looking south by adjusting the altitude of the telescope in the Dob mount. You really need a small motor drive on the altitude axis to do this smoothly enough for guided photography, or a motor-driven wedge under the rocker box. Unguided prime focus photography requires very accurate polar alignment and a carefully adjusted motor speed. Piggyback photography is much less demanding and is a lot of fun! Try that at first. Another method of photography that is fun with a platform is afocal video photography. I made a mount for my 8 mm camcorder to look into a 32 mm Erfle. Turn off the autofocus and manually focus the camera at infinity. I obtained some great shots of Jupiter and the Shoemaker–Levy comet impact sites using this method. I also use an inexpensive, low light level, surplus surveillance camera that has a removable C-mount lens. I use it (with an adapter) at prime focus. Without the lenses in the optical path, and with the camera's increased low light level sensitivity, even some of the brighter deep sky objects are video targets now! Combine the approach using a video camera with a laptop and a small portable frame-grabber like a "Snappy ©"and you have a digital imaging system!

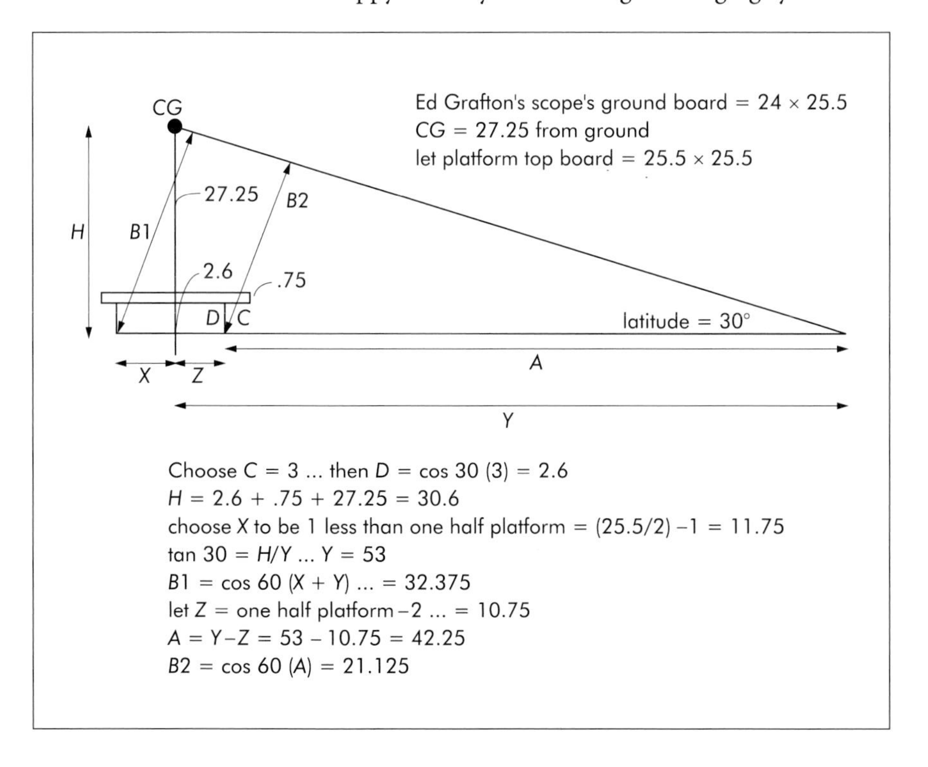

Figure 9.10 Calculating platform dimensions (worked example for latitude = 30°).

Equatorial platforms can provide amazing tracking accuracy, but they are not the same as high-precision German equatorial mounts, but then neither do they have the weight and bulk of a German equatorial. For portable systems such as a Dob-mounted Newtonian, a simple equatorial platform is a very logical tracking system with which to start. Just as with a Dob it is very tempting to add embellishments to supposedly make it better, and thereby depart from the elegant simplicity of the design, so it is with equatorial platforms. The right approach for both Dobs and platforms is to start simple, and only upgrade the system if you need improved performance to meet objectives not currently satisfied, or to satisfy that ever present ATM urge to tinker! Eventually, you can build a more sophisticated tracking system such as a computer-controlled altitude/azimuth system if desired. However, a platform will provide years of enjoyment and increases the versatility of a Dobsonian-mounted Newtonian without detracting from any of the significant advantages of the basic and straightforward design pioneered by John Dobson. Enjoy!

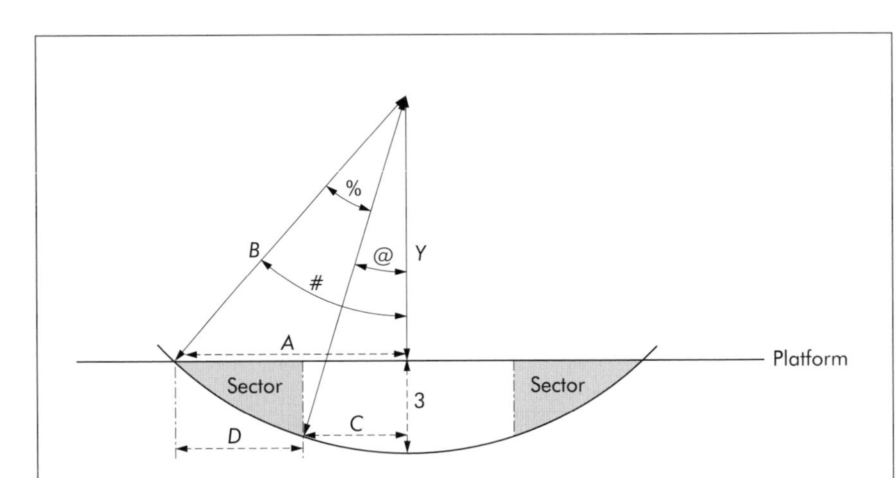

For $B1$ ...

$Y + 3 = B = 32.375$ ... $Y = 29.375$

$\cos \# = Y/B$ ... $\# = 24.86°$

$A = \sin 24.86\ (B) = 13.61$

For 1 hour + travel time ... % = 20°

$@ = \# - \% = 24.86 - 20 = 4.86°$

$\sin @ = C/B$ ... $C = \sin @\ (B)$

$D = A - C = 13.61 - \sin 4.86\ (32.375)$

$D = 10.86$

For $B2$ ...

$Y + 3 = B = 21.125$ ... $Y = 18.125$

$\cos \# = Y/B$ ... $\# = 30.91°$

$A = \sin 30.91\ (B) = 10.85$

For 1 hour + travel time ... % = 20°

$@ = \# - \% = 30.91 - 20 = 10.91°$

$\sin @ = C/B$ ... $C = \sin @\ (B)$

$D = A - C = 10.85 - \sin 10.91\ (21.125)$

$D = 6.85$

Figure 9.11 Calculating sector sizes (worked example from results of Figure 9.10).

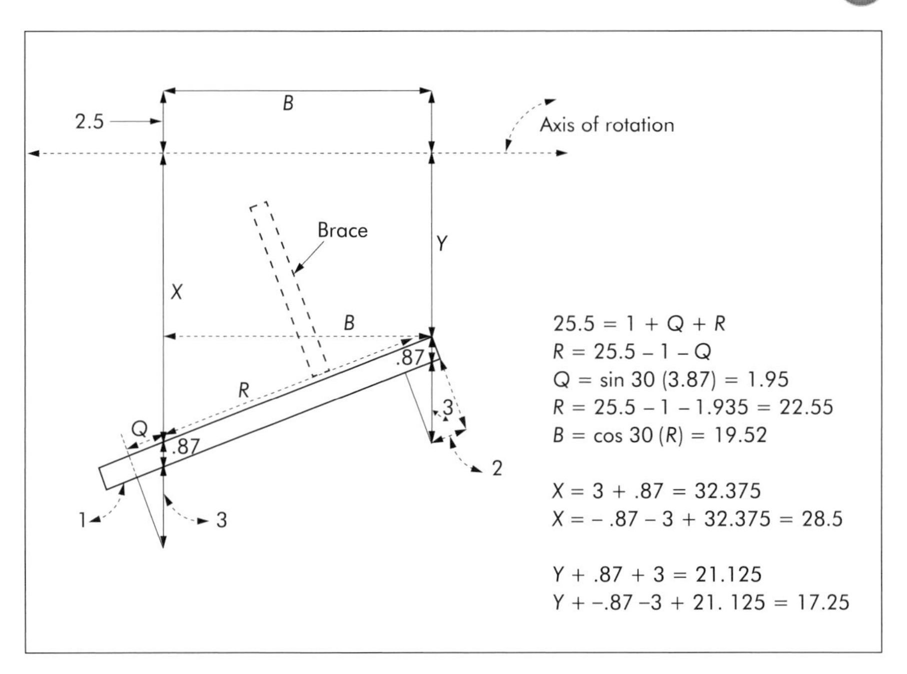

Figure 9.12 Rotation configuration.

Chapter 10
A Computerized Dobsonian

Mel Bartels

One of the characteristics of many ATMs is that, as soon as some new gizmo is available for professional telescopes, they set about finding out how to arrive at the same system in a manner that makes it available to the home constructor. Some of the largest modern professional telescopes use altazimuth mounts. ATMs seized upon this idea as being compatible with the ATM workhorse: the Dobsonian-mounted telescope. Mel Bartels has designed a system whereby the simple Dobsonian can be controlled by a PC and a simple hand-pad. His software will also interface with some planetarium programs in order to give the telescope the GOTO ability of commercially available instruments. This project also exemplifies what is attainable through the principle of mutual aid in the ATM community.

In this chapter, Mel Bartels discusses the principles underlying computer-controlled Dobsonian telescopes. The latest version of his software is available from his web page: <http://zebu.uoregon.edu/~mbartels/altaz/altaz.htm>, which may also be accessed via this book's web page (see p. 249).

Add inexpensive computer-controlled motors to your telescope. If you can build a Dobsonian mount and can solder parts to a printed circuit board, then you have the necessary skills. Plan to spend roughly US$500 for motors, electronics, and gear reducers. Compare this with commercially available computerized mounts for the amateur that cost US $10 000 (Bisque's Paramount).

Imagine a large-mirror Dobsonian telescope equipped with motors. Objects stay centred in the eyepiece – no more constant hand-pushing, no more fear of losing objects when you step away. You find yourself relaxing, enjoying higher magnifications, and seeing more detail.

At f/5, a popular focal ratio of larger aperture telescopes, the coma-free area is only several Jupiter widths across. An object must stay centred in this area to capture those fleeting moments of crystal seeing. Adding motorised focusing means a completely hands-off, vibration free viewing.

A night of perfect seeing at the August 1997 Oregon Star Party illustrates this. Thanks to motorized tracking and precision centring, we were able to run my 20 in (0.5 m) Dobsonian (Figure 10.1) up to 2000×, 100× per inch of aperture, on the Ring Nebula, M57. Much to our surprise, the central star appeared without need of averted vision as a tiny steady speck of light inside a nebulous ring that entirely filled the super-wide-angle eyepiece. Only a nearby 20 in (0.5 m) on a tracking equatorial platform delivered an equal view. Hand-pushed scopes with similar apertures fell behind, unable to use powers above 500×.

Locate objects using point-and-click graphical planetarium programs, or preloaded object files. If star hopping is your game, you can hop from star to star to object, using the motors. Or, you can push the scope by hand, the encoders always keeping track of where the scope is pointing in the sky. At star parties, you can instruct the software to treat hand-pushes of the scope

Figure 10.1 The author's computerized Dobsonian.

as errors, so that if an enthusiastic child accidentally bumps the scope, it will automatically return to the object.

You can put your hands in your pockets, following large extended objects across many high-power fields of view using pre-programmed, smooth scrolling motions. You can initiate outward spiral search patterns. Indeed, any robotic motion that you dream of can be programmed and executed by your computer.

You can vastly extend your reach into the universe by attaching film and CCD cameras to a motorized Dobsonian. Just as the large mirror Dobsonian revolutionised visual amateur astronomy, amateurs with CCD cameras are entering realms formerly held by professionals (Figure 10.2). The confluence of fast, powerful PCs, large mirror Dobsonian telescopes, and CCD cameras propel amateurs into a world only dreamed of a few years ago. Amateurs with CCD cameras already impact fields such as asteroid discovery, supernova discovery, and astrometry. Because of the small chip size, CCD cameras require tight tracking and accurate positioning, exactly what a carefully constructed, computer-controlled mount can deliver.

Figure 10.2 M13 imaged through a computerized Dob.

The Design of a Computerized Telescope

Simple, inexpensive, amateur-built computer-controlled telescopes are still rare because the required design knowledge resides in the hands of only a few amateurs. Knowledge and experience in building large mirror telescopes, motorized drive systems, motor control circuitry, computer hardware, and real-time control software are needed. In addition, PCs with the necessary computing power have been available only recently. Plus, the phenomenal success of the Dobsonian has hidden the limited capability of the Dobsonian design.

System analysis must define the capabilities of the telescope, carefully integrating mechanical and dynamical aspects of the mounting with simple, easy-to-obtain hardware and straightforward software. A simple design consists of a large, thin mirror supported in a sling that is best mounted altazimuthly. A single stepper motor per axis, operating through a roller drive, utilizes software-generated microstepping for tracking and high-speed, overvoltage half-stepping for slewing. Steppers particularly suit altazimuthly mounted telescopes, where infinitely variable drive rates occur. Microstepping control of a stepper motor is produced by software with a technique called pulse width modulation. The motor sees an averaged current at full torque. The PC's timer chip is used to precisely time the stepper's rotor rotation, creating sub-arc-second motion resolution. High-speed slewing is greatly enhanced by a Zener diode flyback circuit in conjunction with temporary overvoltage.

This particular computerized Dobsonian project that I outline calls for a PC to operate the software and control the motors. Any old used 286 AT class machine or better will do. Why not a microcontroller? Using a velocity-based algorithm where a velocity is calculated, then sent to the motor controlling microcontroller, the altazimuth drive rates must be updated 10 to 20 times a second, otherwise the telescope's motion will begin to deviate appreciably from the star's motion. The necessary update rate varies greatly across the sky, but this is a rule of thumb often adopted. The controlling PC must be in tight accurate timing loops, sending precisely timed commands to the microcontroller that in turn, command the motors. Since the PC must be concerned foremost with accurate timing, why not fill the void waiting for the next timing pulse with code that directly generates the microstepping waveforms, thus bypassing the microcontroller completely? Using a positional-based algorithm, where the telescope follows the star from point to point in a line, the positional update need not occur so often, because the deviation from the star's actual arc across the sky is small. Still, the ease of updating PC run software, and sending it out across the Internet that evening to users in time for their evening observing run, cannot be beat.

Consider the advantages of software-based control:

1. Single (any old, small, surplus) stepper motor per axis.
2. Very simple drive circuit.
3. Smooth microstepping tracking. Here is a note from a user showing how accurate the drive can be in a portable scope:

   I went to our dark sky location and took my 14.5 in (0.37 m) with me to do some dark sky imaging. The drive system's performance was breathtaking! I took 21 one-minute exposures of M-101 at f/3.5 (14.5 in f/5 system with a 0.7 focal reducer). When I went to do the track and stack operations, the TOTAL displacement from the first image to the 21st image was only 13 pixels. That is about 30 arcsec over the almost 30 minutes it took to take the images! That's a drift rate of about 1 arcsec per minute of time. Incredible!

4. High-speed, overvoltage, half-step slewing.
5. Low current draw, typically 0.1 A at 12 V d.c..
6. 3 star initialization for more precise pointing.
7. Field derotation.
8. Backlash compensation.
9. Periodic error correction.
10. Refraction correction.
11. Computerized finding using Project Pluto's Guide (planetarium software by Bill Gray), or from a number of contributed data files, or from manual entry of coordinates including offsets.
12. Drift compensation, to track at lunar and solar rates, to follow fast moving comets, and to null tracking rate.
13. Recording of guiding corrections for later analysis and incorporation into periodic correction file.

14. Siderostat option: prevents mount from flipping over and instead moves scope past zenith.
15. Altitude and azimuth software motion limits.
16. Recovery of last position and last orientation to the sky.
17. Move to a home position for blind storage.
18. Optional encoders so that the scope can keep track of its position when moving by hand.
19. Real time display of all coordinates and status.
20. Robotic scrolling motions, best described as high-magnification "flyovers".
21. Grand tour, where a flip of a switch takes you from object to object in a data file.
22. Ability to record scope position from the eyepiece, for later use in data files.

The Motors

Previous state-of-the-art motorized mountings consisted of a slow tracking motor and a high-speed slewing motor, an electrical clutch, and complex circuitry to control each motor. If a motor's ramp up to maximum speed profile needed to be changed, the control circuit often had to be changed. In addition, the constantly varying drive rates required by an altazimuth mount were practically impossible to model in the control circuitry.

Taking advantage of the PCs computing power allows us to directly control a single motor per axis, using a very simple drive circuit. The varying drive rates of an altazimuth mount are easily handled in software, as are different ramp profiles and torque and inertia changes.

D.C. servomotors can more easily be driven over a wide range of speeds than stepper motors. But the positional or velocity feedback loop necessary for precision speed control, obtained either from an encoder or from a tachometer, is very challenging to tune over the range of tracking speeds that an altazimuth mount demands. Backlash and less than perfectly balanced tube assemblies complicate the tuning even more.

D.C. stepper motors are cheap, obtainable in the United States from surplus stores for $2 to $7. Stepper motors run open loop, that is, the software commands the motor to move with no feedback that the motion actually took place. This is not a problem with the modest torque and inertia requirements of even large amateur Dobsonians. In case a motor should stall, the encoders on each axis are used to reset the telescope's position. The real problem with steppers is obtaining smooth motion over a large speed range. Steppers do not like to spin very fast, and if care is not taken, the step–step–step ticking movement of a stepper can be seen at high powers while tracking.

A typical stepper motor consists of a permanently magnetized rotor shaft shaped with radial teeth that rotate inside a stator also containing teeth. Depending on how the stator's teeth are energized, the rotor aligns itself in a particular orientation. The stator has four windings that energise various teeth. To drive a stepper, switch the current from one stator winding to the next.

A full step pattern, or excitation mode, goes like this:

Full step	Winding no.			
#	1	2	3	4
1	ON	off	off	off
2	off	ON	off	off
3	off	off	ON	off
4	off	off	off	ON

At each full step, the rotor aligns itself with the winding that is turned ON.

The half-step pattern, or excitation mode, goes like this:

Half step	Winding no.			
#	1	2	3	4
1	ON	off	off	off
2	ON	ON	off	off
3	off	ON	off	off
4	off	ON	ON	off
5	off	off	ON	off
6	off	off	ON	ON
7	off	off	off	ON
8	ON	off	off	ON

When adjacent windings are ON, the rotor positions itself between the two windings. Steppers move smoothly and are more resistant to resonance effects when half-stepping. Shaft oscillation occurs when the rotor snaps to the next winding during full stepping. The shaft will first overshoot, then undershoot, continuing a decaying oscillation. If the load on the shaft happens to have a harmonic period that matches the rotor's oscillation, a resonance develops between the motor and the load. This can destroy the stepper's ability to rotate at certain rates.

Microstepping

A much bigger improvement in rotor smoothness occurs when microstepping. In the past, amateur altazimuth stepper motor drive designs have sometimes failed because of induced vibration caused by coarse step resolution. With a PC directly controlling the voltage waveform of all four stepper motor windings, we can easily divide each full step into many microsteps.

To microstep: winding A slowing ramps down in current, while winding B slowly ramps up in current. Applying full current to winding A positions the rotor directly over winding A. Applying equal current to both windings A and B positions the rotor directly between windings A and B. Applying current to winding B that is 60% of winding A's current will position the rotor exactly $\frac{1}{4}$ of the way between windings A and B. Owing to the inverse square nature of the electromagnetic force, moving smoothly between windings A and B calls for a particular current pattern to be applied to the two windings.

Limitations on microstepping include absolute tooth error, typically $\frac{1}{25}$ of a full step, and a deflection error caused by torque loading. The deflection error is at a minimum when the rotor is positioned on a winding and at a maximum when positioned between windings. If the torque loading is 10%, then the shaft's error when between windings will be 10% of a full step. Microstepping at 10 microsteps per full step is a reasonable compromise between smoothness and rotor position accuracy. More microsteps can translate into a smoother motion, but will not result in increased rotor position accuracy.

The PC uses the parallel port's 8 bits of output to simultaneously control the current waveform of the eight windings belonging to the two stepper motors. The current waveforms are generated using a technique called pulse width modulation (PWM). Full current is turned ON for a certain time then turned OFF. The cumulative effect of rapidly repeating ONs and OFFs to the motor is the same as if smooth average current was used. By adjusting the percentage of ON vs. OFF the resulting current can be controlled precisely. Torque remains high, whatever the motor's speed, since full current is applied during the ON time.

For adequate current resolution, the sequence of ONs and OFFs will add to 100 or more. For illustration purposes, let us say that the total sequence per phase is 10. If winding A is controlled by bit no. 0 (control word output = 1), and winding B controlled by bit no. 1 (control word output = 2) of the control word, then the sequence of control words for a single full step with maximum average current (ignoring the other windings on bits no. 2 through no. 7) is:

Sequence of control words output (10 pulses per phase)	
Phase 1	1 1 1 1 1 1 1 1 1 1
Phase 2	2 2 2 2 2 2 2 2 2 2

For full stepping at half-current:

Sequence of control words output (10 pulses per phase)	
Phase 1	1 1 1 1 1 0 0 0 0 0
Phase 2	2 2 2 2 2 0 0 0 0 0

For half-stepping at half current where the intermediate half-step consists of both winding A and winding B on:

Sequence of control words output (10 pulses per phase)	
Phase 1	1 1 1 1 1 0 0 0 0 0
Phase 2	3 3 3 3 3 0 0 0 0 0
Phase 3	2 2 2 2 2 0 0 0 0 0

To microstep, we place the rotor at intermediate positions between windings A and B. To set the rotor one-fourth of the way towards winding B, the rotor must "feel" winding B one-third as much, positioning itself three times closer to winding A than winding B. Since electromagnetic fields propagate as the inverse square, the current supplied to winding B must be sqr (1/3), or about 60% of current to winding A:

Sequence of control words output (10 pulses per phase)	
Winding A at 100% current	1 1 1 1 1 1 1 1 1 1
+ Winding B at 60% current	2 2 2 2 2 2 0 0 0 0
= Winding A + Winding B	3 3 3 3 3 3 1 1 1 1

Therefore, to microstep with four microsteps per full step with maximum current:

Sequence of control words output (10 pulses per phase)		
Phase 1:	1 1 1 1 1 1 1 1 1 1	(A current = 100%, B current = 0%)
Phase 2:	3 3 3 3 3 3 1 1 1 1	(A current = 100%, B current = 60%)
Phase 3:	3 3 3 3 3 3 3 3 3 3	(A current = 100%, B current = 100%)
Phase 4:	3 3 3 3 3 3 2 2 2 2	(A current = 60%, B current = 100%)
Phase 5:	2 2 2 2 2 2 2 2 2 2	(A current = 0%, B current = 100%)

For 10 microsteps		
Phase 1:	1 1 1 1 1 1 1 1 1 1	(rotor positioned on winding A, A = 100%, B = 0%)
Phase 2:	3 3 3 1 1 1 1 1 1 1	(rotor positioned 9:1 or 9 times closer to A, A = 100%, B = sqr (1/9) = 33%)
Phase 3:	3 3 3 3 3 1 1 1 1 1	(rotor positioned 8:2 or 4 times closer to A, B = sqr (1/4) = 50%)
Phase 4:	3 3 3 3 3 3 3 1 1 1	(rotor positioned 7:3 or 2.3 times closer to A, A = 100%, B = sqr (3/7) = 65%)
Phase 5:	3 3 3 3 3 3 3 3 1 1	(rotor positioned 6:4 or 1.5 times closer to A, A = 100%, B = sqr (2/3) = 82%)
Phase 6:	3 3 3 3 3 3 3 3 3 3	(rotor positioned 5:5 or equal distance from A and B, A = 100%, B = 100%)
Phase 7:	3 3 3 3 3 3 3 3 2 2	(opposite of Phase 5)
Phase 8:	3 3 3 3 3 3 3 2 2 2	(opposite of Phase 4)
Phase 9:	3 3 3 3 3 2 2 2 2 2	(opposite of Phase 3)
Phase 10:	3 3 3 2 2 2 2 2 2 2	(opposite of Phase 2)

These ten values are defined in the software, written in C, as an array: PWM[0] through PWM[9]. In C, the first element of the array has an index or offset of 0.

Slight tweaking of the PWM values is necessary to reflect the finite on-off times of the power transistors, hex inverters, any opto-isolators used, the parallel port, differences in speed between PCs, and differences between motors and the torque loading.

High-Speed Overvoltage Half-Step Slewing

Besides excessive vibration when full stepping, stepper motors have another limitation to overcome: they do not like to spin very fast. As the computer switches current ON and OFF to the windings, counter electromotive force (e.m.f.) is generated. When the source of the current is switched OFF, the collapsing magnetic field quickly moving through the winding generates a voltage spike that can destroy the power transistors.

A flyback diode prevents the voltage spikes by giving a path for the dying current to circulate back into the winding. However, this greatly slows the time for the current to collapse. The result is ever lowering torque as the motor tries to spin faster. A Zener diode used with the flyback diodes allows just the voltage above the Zener diode's rating to be returned to the power source. This prevents the extreme voltage spiking while avoiding the full braking action of the flyback diodes.

In combination with using higher voltage than the motor's continuous voltage rating, and smoothly ramping up the motor's spin, we can achieve speeds many times faster than otherwise. Rates up to 5000 half-steps per second can be achieved with modest torque. I use two 12 V batteries in series to generate a total of 24 V to operate 6 V steppers. This gives enough voltage to run the steppers at a high speed. A single 12 V battery also operates the steppers adequately. Current consumption for both motors combined is 0.1 A while microstepping and 0.3 A while slewing.

A circuit diagram for the computerized Dob is given in Figure 10.3 (*overleaf*). A printed circuit board is available – contact the author for details.

PC Parallel Port

The parallel port is an ideal interface for controlling telescopes, particularly with laptops in the field. The

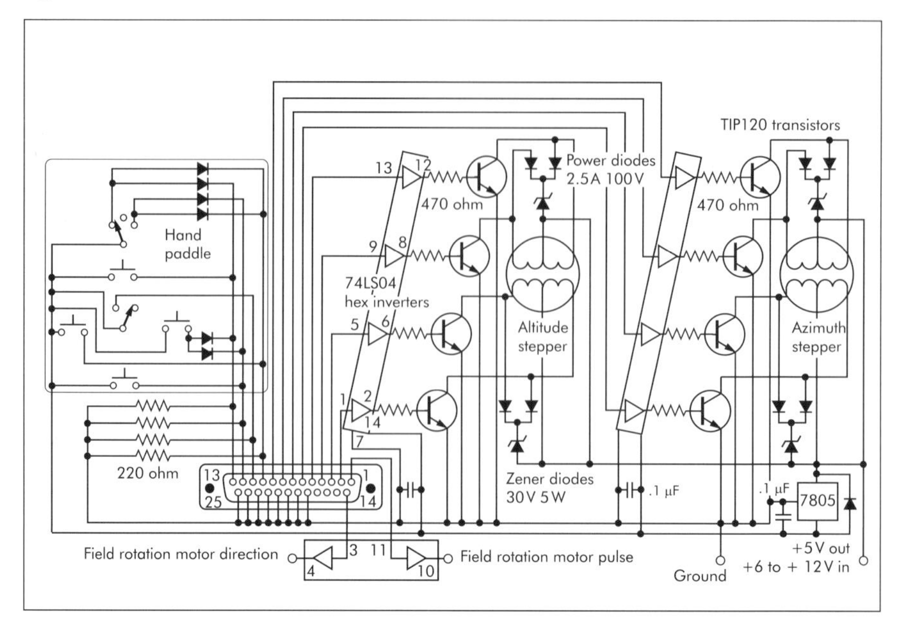

Figure 10.3 Circuit diagram for the computerized Dob.

parallel port uses 8 bits of output, typically at port address 0378. On the 25-pin connector, the 8 bits of output are on pins 2 through 9, from least significant to the most significant bit. These 8 bits of output are perfect for controlling the two stepper motors needed to drive a telescope in altazimuth mode. The parallel port cannot provide or sink large currents directly, hence, 74LS04 hex inverters are used to interface between the parallel port and the driver transistors.

The parallel port has 5 bits of input at port address 0379. These input bits are on pins 15, 13, 12, 10, and 11of the 25-pin connector, with 11 being inverted. Pin 15 activates bit 8, pin 13 activates bit 16, pin 12 activates bit 32, pin 10 activates bit 64 and pin 11 activates bit 128. Depending on which lines are raised high, the values can range from 8 through 248. In addition, the parallel port has 4 bits that can be either in or out at port address 037A. These 4 bits show up on pins 1, 14, 16 and 17 of the 25-pin connector. Bits 1,14 and 17 are inverted. These values when read range from 0 to 15.

Mechanics

We want to set the stepper motor step size as a compromise between microstepping tracking resolution and a fast slew rate. In addition, the tracking step size and rate must be small and fast enough to not resonate with the telescope's natural harmonic frequency. A microstep size of $\frac{1}{4}$ arcsec is smaller than the typical atmospheric resolution limit of 2 arcsec, and is tiny enough to not cause the telescope to jitter. Most amateur scopes resonate at several hertz, that is, if you tap the tube while looking through the eyepiece, you'll see several vibrations per second that take a second to die off. So, a rate of 30 microsteps per second is quick enough to avoid causing the telescope to vibrate. If we take the sidereal tracking rate of 15 arcsec per second of time as an example, then the microstep size can be no larger than $\frac{1}{2}$ arcsec. Through the eyepiece, the telescope tracks smoothly: it does not jitter from microstep to microstep.

If we ramp the steppers up in speed, we can achieve 5000 half-steps per second. If we adopt a microstep size of $\frac{1}{4}$ arcsec to $\frac{1}{2}$ arcsec where a full step is divided into 10 microsteps then the top slew speed is from $1\frac{3}{4}°$ to $3\frac{1}{2}°$ per second. This is plenty fast to move a large scope and give time to duck!

Most stepper motors have 200 full steps per revolution. The reduction needed between motor and telescope is 360°, divided by the distance one stepper revolution covers. If there is $\frac{1}{4}$ to $\frac{1}{2}$ arcsec per microstep, and 10 microsteps per full step, and 200 full steps per revolution, then one stepper revolution covers 500 to 1000 arcsec. Dividing this into 360° or 1 296 000 arcsec calls for a reduction of 1300:1 to 2600:1 between motor and telescope.

Taking advantage of gravity, we can place the large altitude bearing on a small driveshaft of $\frac{3}{8}$ to 1 in (1 to 2.5 cm) diameter. Make the altitude bearing from wood using a router on a extension bar anchored by a pivot. This makes an extremely accurate rim. Face it with a thin aluminium strip. Large Dobsonians have altitude bearing diameters of 12 to 24 in (30 to 60 cm). This gives a reduction of 12:1 to 64:1, depending on drive shaft and altitude bearing diameter selection. Remembering our required 1300:1 to 2600:1 reduction between motor and telescope, we have a further 20:1 to 220:1 to cover. Small instrument gear reducers are the most popular choice to cover this remaining reduction. A timing belt on pulleys, and a roller reducer with a small metal shaft and a larger metal disc, are two other choices. See Figures 10.4 and 10.5 (*overleaf*).

The gear reducers can have relatively large periodic error. If the gear reducer periodic error is 2 arcmin, and the reduction is 60:1, then the periodic error at the eyepiece will be 2 arcsec.

Figure 10.4 Baseboard and azimuth drive.

Figure 10.5 Altitude drive, viewed from underneath the baseboard.

One other drive scheme deserves mention. A threaded bolt from 4 × 40 to $\frac{3}{8}$ in × 16 threads per inch (M3 to M10) is pressed into threads formed on a rim. Because several threads are in contact at any one time, errors are averaged out. The threads can be formed from wood putty, epoxy, nylon, brass, or from two threaded bolts bent around the rim and placed side by side. A $\frac{1}{4}$ in × 20 tpi (M6) drive on a rim of 20 in (50 cm) diameter yields approximately $\frac{1}{2}$ arcsec microsteps. Periodic error of several arc seconds typically results from this type and size of drive.

Encoders

As long as the stepper motors never stall, as long as the scope is not bumped, the software always knows where the scope is pointing. This state of affairs does not last forever, particularly at public star parties! After the scope is bumped, not only is the computerized finding broken, but also the tracking is bad, because the drive rates vary for each position in the sky. You must centre the scope on a known object, and inform the controlling software what object the scope is pointed at.

The recourse to this is to use encoders. These small (2 in or 5 cm diameter), lightweight (a couple ounces or 50 g), inexpensive (US$50) devices convert rotary motion into digital pulses. These pulses are counted by a microprocessor. The controlling PC/laptop queries the microprocessor via a RS232 serial connection for the current counts, converting the counts to shaft angles. The parallel port is engaged in sending output to the stepper motors and receiving control signals from the hand paddle, and the serial port is engaged in communicating with the encoder interface.

The popular incremental optical shaft encoder consists of an optical disc with alternating clear and opaque spokes. Two LED light sources shine onto detectors through the spokes. The LEDs are slightly staggered such that when the optical disc is rotated, the following sequence occurs (actually, a number of LED pairs may be employed):

```
                    Time --->
Outer detector      ON   off   ON   off
Inner detector         ON   off   ON   off
```

If the disc is rotated the opposite direction, then the sequence occurs backwards, and the inner detector turns ON before the outer detector:

	Time --->							
Outer detector		ON		off		ON		off
Inner detector	ON		off		ON		off	

The microprocessor decodes each passing of a spoke into four events. This is called quadrature decoding. If an optical disc has 2048 spokes, then the microprocessor quad decodes this into 8192 counts per revolution.

The microprocessor and its software must be able to handle the speed of the pulse train from the encoder. If the encoder is geared 8:1, giving 64,000 counts per revolution, and the shaft is spun at one revolution per second, then the total events from the outer detector, called the "A" channel, and from the inner detector, called the "B" channel, occur at the rate of 64,000/s. To reject noise, the processor should sample each channel three times, accepting the result only if all three reads are the same. Finally, the processor has to continue counting while handling communications with the controlling PC/laptop.

In a permanently mounted telescope, the encoder interface need never be powered off, and can be maintained by a small battery. When the controlling PC/laptop is shut down, the encoder interface will continue to count encoder pulses, always knowing where the scope is pointed. In an altazimuth mount, only the tube assembly in altitude need be set to a known angle at program start up time. The popular Taki routine, used as the basis for translating altazimuth to equatorial coordinates in many software packages, does not need to synchronize the starting azimuth value to the scope's starting direction.

One way to set the altitude angle to 0° is to set up the telescope, paying no particular attention to the base being exactly parallel to the ground. Point a precision bubble level in different directions, finding the direction on the base that is exactly level. Rotate the scope in azimuth until the tube points the same direction. Level the tube, thus aligning it parallel to the base. This presupposes that the optical axis exactly parallels the tube's mechanical axis.

Alternatively, you can set the altitude to 90° elevation by using a stop consisting of a threaded bolt run through the back of the Dobsonian rocker. Adjust the threaded bolt ahead of time using the following method suggested by Richard Berry: draw a reticle pattern on a piece of paper and exactly centre the paper over the focuser with the eyepiece removed. Point the scope upward. Aim a camcorder bolted to a rigid framework down into the tube assembly just to one side of the diagonal and focus on the paper reticle. Turn the tube assembly in azimuth, adjusting the threaded bolt and always pushing the tube assembly up against the bolt, until the reticle pattern stops looping. If the azimuth bearing is not exactly flat, this will be impossible to achieve. One side of the Dobsonian rocker may have to be adjusted slightly higher or lower in order to get the reticle to spin in place.

A good encoder model is US Digital's (3800 NE 68th Street, Suite A3, Vancouver, WA 98661-1353, USA, Voice (360) 696-2468, Sales (800) 736-0194, Fax (360) 696-2469) S2-2048-B, quad decodes to 8192 counts per revolution; the price at time of writing was US$66 (£45).

David Lane, author of The Earth Centred Universe (ECU), has designed an inexpensive serial interface for quadrature encoders, called the MicroGuider (MGIII). Cost for parts, not including the encoders, is about US$100. Bob Segrest has designed a very small PCB. Patrick Dufour offers a very compact commercial encoder interface box. Reach him at Ouranos USA, 411 Whitman St., Hanson, MA 02341, Tel. 781-447-2744, or outside the USA at Ouranos main office, 189, rue Du Foin, Saint-Augustin de DesMaures (Quebec), Canada G3A 2S6, Tel. 418-878-9426. A number of vendors make quadrature encoder interfacing chips that count pulses and handle noise rejection. Hewlett-Packard's HCTL-2016 from their Opto-Electronics catalogue is one example.

Field Derotation

In any scope where an axis does not point exactly to the celestial pole, a slow rotation of the field of view occurs. Field rotation is independent of the size of the field of view; for instance, a large scope at high power will experience the same angle of field rotation as a small rich field telescope at low power. Field rotation varies greatly over the sky, non-existent in the east and west, terrible at the zenith, and approximating the sidereal tracking rate when the scope is pointed at the meridian.

This project includes circuitry and software so that a motor can be added to slowly rotate the focuser to compensate for field rotation. Since the field derotator unit need only rotate slowly with finite steps, a simple single driver chip, the SAA1042, is all that is needed, with step and direction inputs. To keep vibration minimised, use a field rotation per step size smaller than 1 arcmin. This is fine enough to prevent field rotation from showing on fine-grained 35 mm film. See Figure 10.6.

For visual imaging, a field derotator motor is not needed. For CCD imaging, exposures can range up to 5 to 20 minutes in many sections of the sky with typical chips, before field derotation is needed. If you have a CCD chip with very large numbers of pixels, or are planning to do prime-focus astrophotography, then you will want to add the field derotator motor.

Field rotation is very insensitive to errors in aligning the telescope to the sky. A large telescope-to-sky misalignment of $\frac{1}{2}°$ results in only a very tiny fraction of a degree difference in field rotation angle, not great enough to appear at the edge of a CCD detector or the edge of a film frame.

The control program shows the field rotation in real time, so that you can plan exposures without a field derotator. Here is how I judge field rotation for a CCD chip that is several hundred pixels on a side. The total number of pixels on the perimeter is roughly 1000. That means I can tolerate a field rotation of 360° per 1000 resolution units, or about $\frac{1}{2}°$. I watch the scope

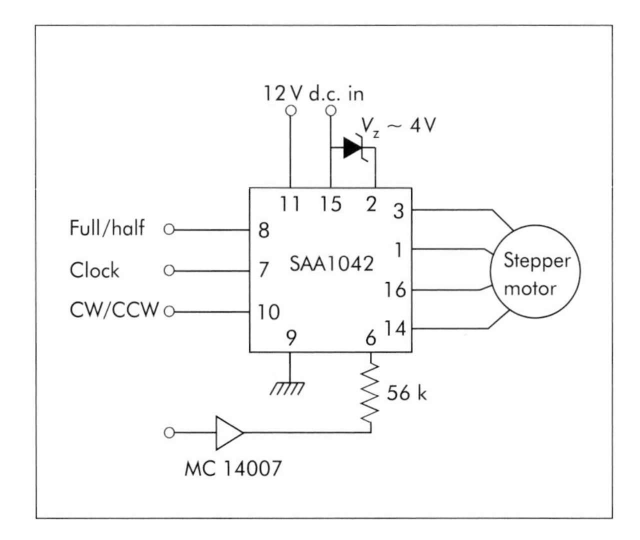

Figure 10.6 Field derotator circuit.

track in real-time, noting the amount of field rotation change over 10 s or so. If it looks like I can image for my desired exposure time, then I go ahead. If not, I wait until the object is better positioned in the sky, or I shorten my exposure time.

Pointing Errors

Several pointing errors that can be handled very nicely in software are backlash, periodic error, residual drift, and refraction.

Backlash

Handling backlash is more complicated in an altazimuth mount than in a traditional equatorial mount. In an equatorial mount, one can reverse the tracking motor for a distance at least equal to the backlash amount, then move the motor forward the same amount. This will always take up the backlash because the tracking motor tracks in one direction only. In an altazimuth mount, the motors will track in different directions, depending on where the scope is pointed. My algorithm keeps track of the amount of backlash that needs to be taken up in each direction. If the motor changes direction, and continues to move in this new direction several times, then the motor is quickly moved to cover the backlash that needs to be taken out. It's important that the motor direction stays constant over several moves or else the motor may oscillate back and forth as the backlash is taken out first in one direction then in the other. Consider the following situation where the scope is at the meridian and the change in altitude is virtually non-existent. Eventually, the change in altitude accumulates to the point where a single microstep command to the altitude motor can be sent. This occurs when the accumulated altitude change reaches the distance equal to half a microstep. All motions are averaged, that is, if a movement of 32.3 microsteps is needed then a command to move 32 microsteps is sent, and if the movement required is 32.8 then the command is to move 33 microsteps. After the motor moves the scope, the altitude movement needed is actually in the opposite direction, but less than half a microstep, so no motion occurs. However, the backlash-checking routine only knows that the motor needs to move back in the opposite direction, and incorrectly attempts to take out the backlash in this opposite direction. By requiring that the motor move several times in the same direction, this oscillating backlash is avoided. Eventually, the change needed in altitude points the motor back to the original direction, and another single microstep command is sent.

Periodic Error Correction

Many modern professional scopes use large circular rollers driven by machined shafts. These avoid the errors inherent in worm-and-gear drives. Worm-and-gear include periodic and erratic errors. Periodic errors are caused by the elliptical shape of the gear and by mis-centring of the worm on its shaft. Erratic errors are caused by tooth to tooth differences and by backlash when the drive changes direction.

A periodic error shows itself by a slow oscillation of a star back and forth across a high-power guiding eyepiece reticle. If one is using a single-turn 60:1 gear reducer on a 200-full-step-per-revolution motor, then the periodic error repeats every 200 full steps, since the worm is usually responsible for the majority of error in a gear set. If using a quad turn worm, then the periodic error will occur over 800 full steps. If the 60:1 gear reducer is attached to another gear reducer, then there will be two errors with periods of 200 full steps, and 12 000 full steps.

Periodic error correction (PEC) is handled in both axes simultaneously, covering any number of desired full steps. The software includes a guiding mode, where guide corrections related to the motors' shaft angle are saved to a file for later analysis.

Generate several files and import them into a spreadsheet.

Using the motors' shaft angle expressed in full steps, align the data from each file on top of each other using spreadsheet software. Then average the data. Remove residual drift where the start of the curve is higher or lower than the end of the curve. Select a single period out of the data. The resulting numbers from each axis are used to generate a periodic error correction file (PEC.DAT). Remember that the motors must start at a predetermined angle that corresponds to the start of the PEC.DAT file. If the motors stall at all, they must be resynchronised to the start of the PEC.DAT file. In some cases, the gear reducer will have such steep and fast-rising periodic error that it is practically impossible to synchronize the motors to the PEC data. The only recourse is to try another gear reducer. A thorough and clear example of actual periodic error reduction by Joe Garlitz can be found on the web (Garlitz, 1997).

Drift

After backlash and periodic error are removed, portable telescopes show some residual drift, typically a couple of arc seconds per minute. This is due to slight inaccuracies in the exact centering of the initialization stars, and slight inaccuracies in the mount itself. To fix this, centre a guide star and start the guide function. After a minute, recentre the star and end the guide function. The software will automatically calculate the drift in altitude and azimuth, display it, and use it immediately. You will have to do this whenever the scope is moved to a new section of the sky. The result is unguided tracking accurate to an arc second or so from 1 to 5 minutes' time. You should be able to get unguided CCD images from 30 s to several minutes in duration.

Refraction

Refraction is a lensing effect of the atmosphere where the light from distant astronomical objects near the horizon is bent upward by the very long path it must take through the atmosphere. The bending acts to make objects appear higher than they really are, and to require a slower tracking rate. An averaged value of refraction is easily calculated on the fly by software and compensated for.

Software

The software has gone through several reincarnations, starting as 6502 assembly code for the Commodore 64, when the Commodore 64 first came out. (I bought my

Commodore 64 for an at the time incredible sales price of $600, and with no floppy or tape drive, and had to re-enter my programs every time I turned the computer back on!) Unfortunately, the stalwart 2 MHz 6502 processor could only muster recentering the object every couple of seconds. The dream of an inexpensive, amateur-built altaz drive seemed far away, until the AT class machines arrived. The code was then rewritten in C. Later, the code went through its C++ object-oriented life on a 386. Now, in the interests of making the code as universal and easy to port, the code lives in ANSI C. Functions that are directly tied to low-level hardware such as the parallel port and bios clock, use pointers to access the appropriate memory locations. For non-DOS machines, some modification of these parts of the code will be necessary.

The program is based on the popular two-star conversion algorithm, based on an article by Toshimi Taki (Taki, 1989), to translate between altazimuth and equatorial coordinates. The scope need only be accurately aligned on two widely separated stars using a high-power reticle eyepiece; there is no need to level the base. The scope can also be initially set on a planet, say, soon after sunset. After a couple of minutes of microstepping recentering, the scope is initialised on the same object again. The scope will continue to track the object, keeping it in the eyepiece field of view for an hour or two.

In addition, the program will use a third initialization point, for more accuracy than the two-star initialization would give otherwise. Any of the three initialization positions can be reinitialized as often as wanted. The conversion algorithm allows the input of mount construction errors. For instance, one altitude bearing may be a bit lower than its counterpart. Normally this would cause a pointing error, but the conversion algorithm will compensate once given the amount of the error. All initialization positions are saved to a file for later analysis.

The software is event driven by either keyboard or hand paddle input. If no events occur, then the scope moves to the current equatorial coordinates. If the coordinates remain unchanged, the scope tracks. If new coordinates are entered, the scope slews.

I use the PC's bios clock tick as the timer for my software. In detail, the sequence of events for each bios clock tick (which occurs about 18.2 times a second) is:

1. add equatorial drift to current equatorial position,
2. update a status field or work with the optional encoders: either a calculation or a direct write to video memory or reading the encoders or setting current coordinates to encoder coordinates,
3. check for keyboard event, if none,
4. check for hand paddle event, if none,
5. check for IACA event, if none,
6. check for LX200 event, if none,
7. check to see if field rotation motor needs pulsing,
8. then move to current equatorial coordinates by the following steps:
9. calculate new altazimuth coordinates based on new sidereal time that was calculated when bios clock tick occurred,
10. find difference between current altazimuth coordinates and newly calculated altazimuth coordinates,
11. find distances to move in each axis and decide between microstepping and half-stepping, if microstepping, then check for backlash, if none,
12. then spread microsteps over the bios clock tick by dividing number of microsteps into MsTicksRep, the count of PWMs per bios clock tick: if microsteps exceeds MsTicksRep, then reduce number of microsteps per full step up to half-step,
13. continuously generate PWMs, checking for bios clock tick at end of each PWM: a PWM consists of outputting to parallel port an already calculated array of ons and offs to the stepper motors' windings,
14. when bios clock tick occurs, PWMs end and new sidereal time is calculated,
15. current altazimuth coordinates updated to reflect number of microsteps that actually occurred,
16. current altitude coordinate updated to include refraction,
17. current altazimuth coordinates updated to include any backlash compensations already moved,
18. current altazimuth coordinates updated to include PEC based on steppers rotors' position,
19. current altazimuth coordinates updated to include altazimuth drift,
20. current altazimuth coordinates updated to include any guiding motions.

The Impact of the Internet

When I was a beginner in amateur astronomy many years ago, several amateurs including Bob Kestner gave freely of their time and advice. I wanted to repay their efforts; consequently I placed this project on computer bulletin boards in 1992 and eventually the Internet, offering it freely in the best tradition of amateur astronomy and the Internet. Since then, I have been very pleased with unsolicited contributions from others, making this project far better than I could ever achieve alone. Richard Berry suggested using three stars instead of two stars for the initialization, Tom Cathey debugged encoder routines, Dale Eason offered software improvements, Berthold Hamburger and Pat Sweeney developed printed circuit boards, and Chuck Shaw developed the field derotation unit and wrote configuration documentation. Because this project is based almost entirely on software, almost all requested enhancements are answered, the new software going out immediately over the Internet.

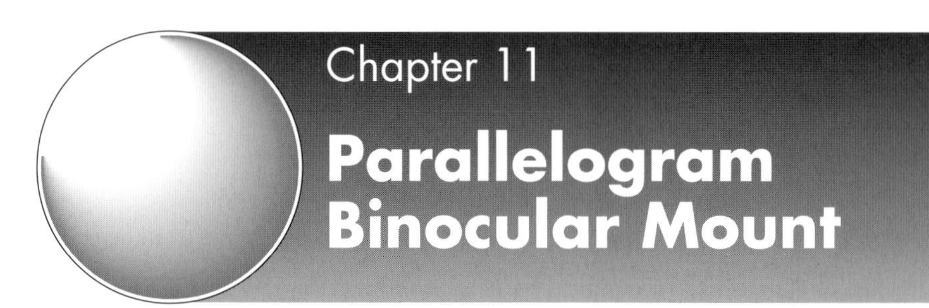

Chapter 11
Parallelogram Binocular Mount

Scott Wilson

Binoculars are awkward to use from a tripod when viewing objects of high elevation. This parallelogram mount holds the binocular away from the tripod and allows its height to be changed without altering its orientation enabling, for example, adults to show youngsters objects of interest. Self-construction of a mount like this enables the ATM to design it from scratch and suit it entirely to his needs, as Scott Wilson shows in this integrated mount and tripod which he has made so that its setup and use is as convenient as possible.

For many years after I bought my 11 × 80 giant binoculars in 1986 for Comet Halley, I mounted them on an ordinary camera tripod. This worked, although it meant getting down on my knees and craning my neck in all sorts of awkward positions, which my guests had to do also. I was then inspired by a commercial binocular mount called the Grandview in a review of binocular mounts in the July 1993 issue of *Sky & Telescope*. I replaced the metal universal joints with simpler wood designs and added a separate counterweight for the binoculars. This design was later modified to its present form by myself and Robert Maxwell as part of an ATM project of the Midlands Astronomy Club in Columbia, South Carolina. This mount is designed to hold 11 × 80 giant binoculars, although it can be adapted for other sizes, or perhaps a small telescope. Its major advantage is that you can stand comfortably and simply pull the binoculars to your eyes. It also adjusts in height for adults and children, and stays pointed at the same object in the sky throughout its vertical range of motion.

Having this mount has increased my binocular observing ten-fold. The only skills it requires, other than basic cutting, drilling and finishing, is the use of a hole-saw. You will also need an L-shaped bracket, which is usually included with large binoculars or can be purchased separately. This bracket attaches to a threaded hole in the central shaft of the binoculars. The mount breaks down into three convenient pieces for carrying, and all fasteners are bound except for a single wing-nut.

You can use any hardwood. I have had success with oak, maple and poplar. You can use pine as long as it's not too soft. You can also substitute commonly available woodscrews for the cabinet screws.

The pivot box (Figure 11.1) is where you attach the binoculars, using the L-shaped bracket. The bracket, which is threaded, is attached with a single bolt through the $\frac{1}{4}$ in (6 mm) slot. The cores from a $1\frac{1}{4}$ in (32 mm) hole-saw are used as rests and contact the swing arm when it is moved to its lowest possible vertical position. You can use dowel slices also. They can be glued in place, if desired, but make sure to align them with a piece of threaded rod. The length of threaded rod is flexible: for my 11 × 80s, I used 12 in. Since the hole in the weight is $1\frac{1}{8}$ in (30 mm) in diameter, I used a $\frac{1}{2}$ in (12 mm) slice of $1\frac{1}{8}$ in (30 mm) dowel as a spacer to hold it in place on the rod. To adjust the weight, you simply loosen the wing-nuts and move it up or down the threaded rod.

Once you've made the pivot box, you need to position the weight on the rod so that it balances the binoculars. After attaching the binocs, support the pivot box through its central hole on a threaded rod or by some strong string, or after installing it in position on the head assembly, which is described below. Find the position of the counterweight on the threaded rod which balances the binoculars. Next, you have to place the L-bracket in its slot in order to move the centre of mass as close as possible to the central hole axis. See if, when positioned at 45° , the box wants to tip one way or the other. If it wants to rotate, change the bracket position in the slot and try again. Eventually it should be stable horizontally, vertically, and at all intermediate positions. The balance need not be perfect, however, since there will be some friction in the pivot box attachment.

The head assembly holds the pivot box on the end of its tongue, labelled ABA in Figure 11.2 (*p. 182*). The

Figure 11.1 The pivot box.

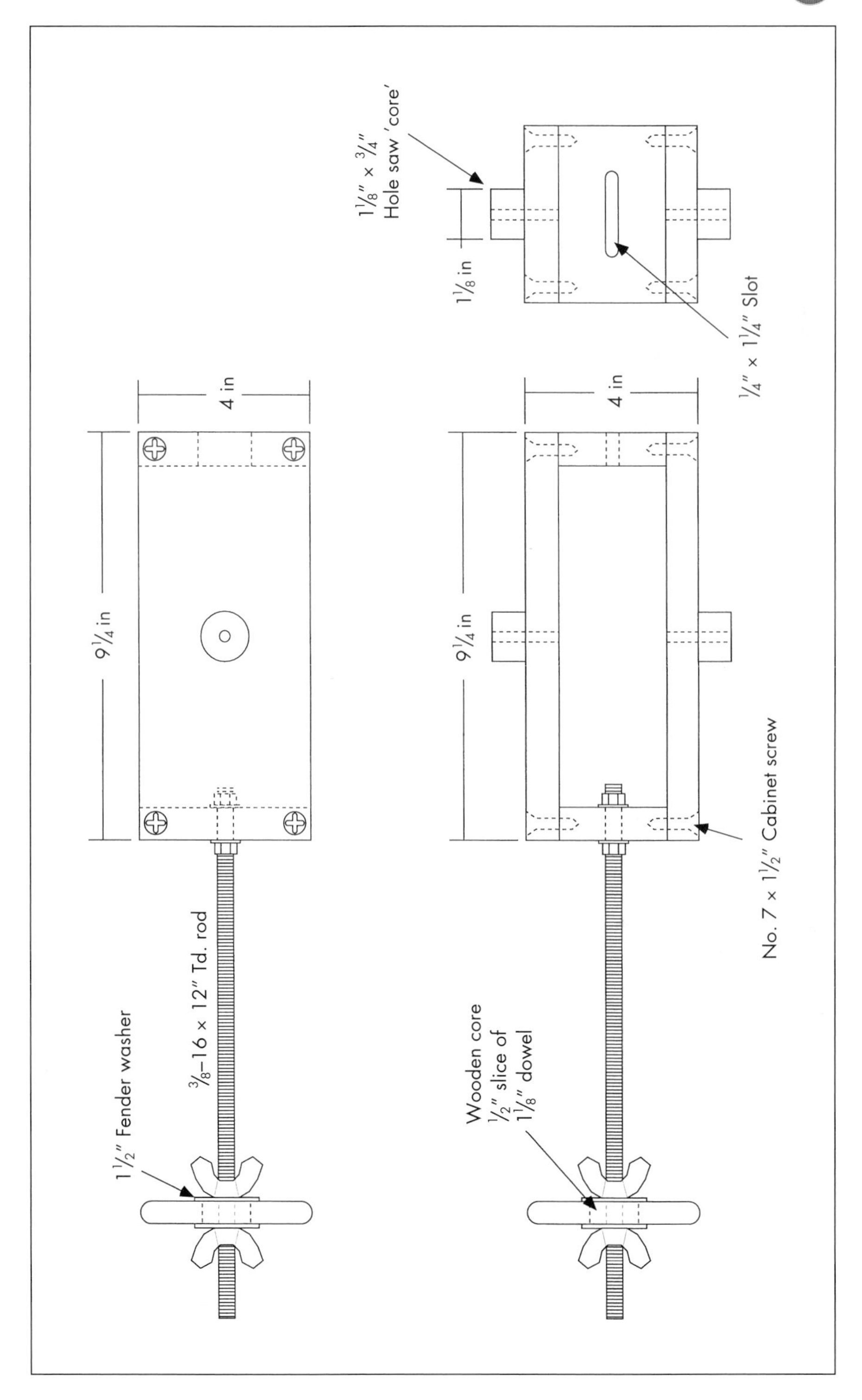
1⅛″ × ¾″
Hole saw 'core'
1⅛ in
¼″ × 1¼″ Slot
4 in
9¼ in
4 in
9¼ in
No. 7 × 1½″ Cabinet screw
1½″ Fender washer
⅜–16 × 12″ Td. rod
Wooden core
½″ slice of
1⅛″ dowel

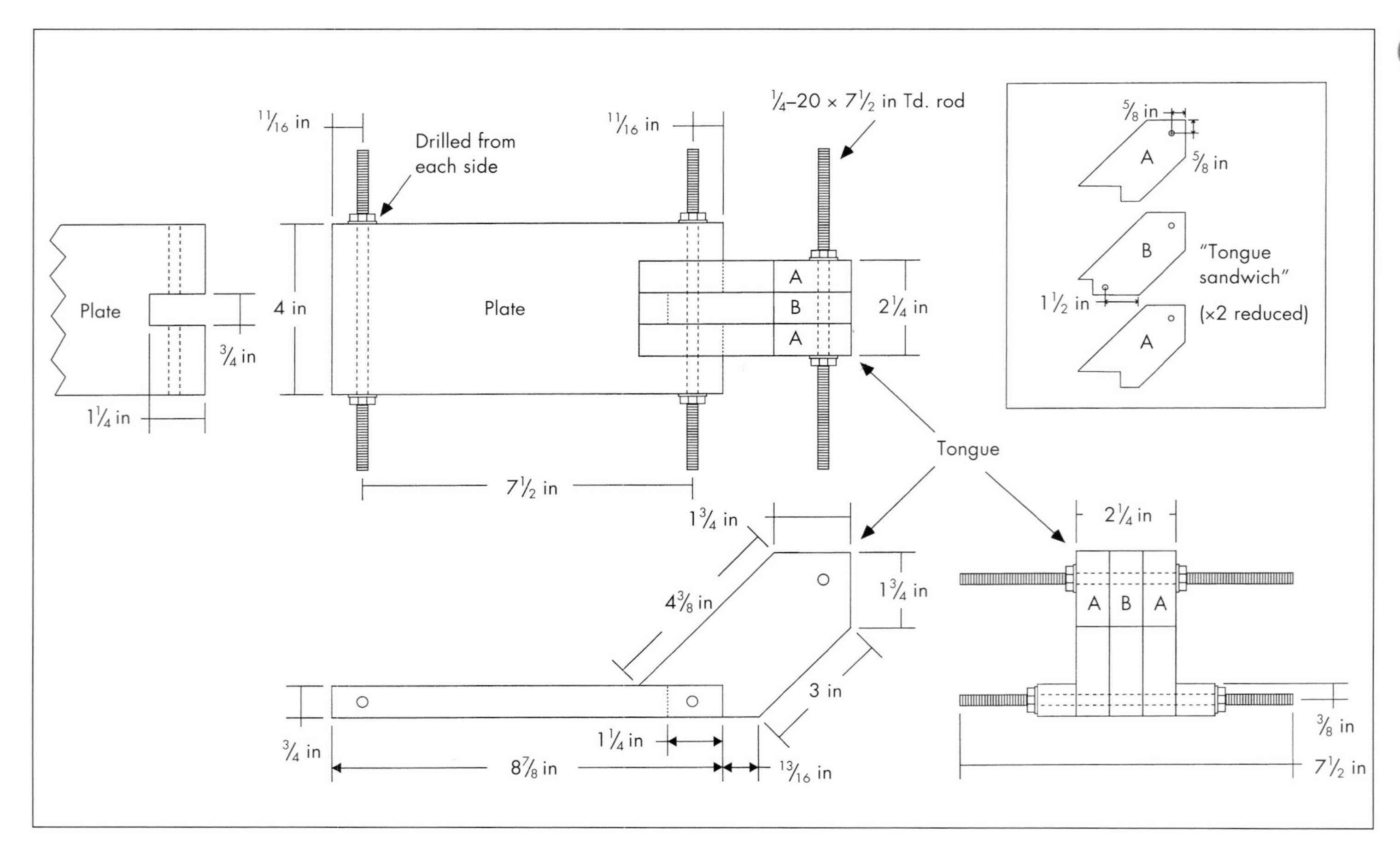
11/16 in
Drilled from each side
11/16 in
1/4–20 × 7 1/2 in Td. rod
Plate
3/4 in
1 1/4 in
4 in
Plate
A
B
A
2 1/4 in
7 1/2 in
5/8 in
A
5/8 in
B
"Tongue sandwich"
1 1/2 in
(×2 reduced)
A
Tongue
1 3/4 in
2 1/4 in
4 3/8 in
1 3/4 in
A
B
A
3 in
3/4 in
1 1/4 in
8 7/8 in
13/16 in
3/8 in
7 1/2 in

Figure 11.2 The head assembly.

layers should be glued for strength. When drilling the plate part, I found it easiest to drill from each side and meet in the middle. It is important to preserve the same $7\frac{1}{2}$ in (19 cm) spacing here that you have on the central pivot, since this is what forms the parallelogram and keeps the binoculars pointed at the same object when you move the swing arm up and down. If you try to drill all the way through from one side to the other, the drill bit can wander and come out in the wrong location. The tongue should be clamped in position when you drill the top of the plate in order to ensure a tight fit of its "B" layer in the plates slot.

Between the inside of the pivot box and the tongue, there are six $\frac{1}{4}$ in × 1 in (6 mm × 25 mm) stainless fender (mudguard) washers, three on each side, which act as spacers. You may have to use more or fewer washers depending on their thickness. When you install the pivot box, align it on the tongue while carefully holding the washers in place, and insert the threaded rod. On the outside of the hole saw cores, I put a $\frac{1}{4}$ in (6 mm) flat washer, a wing-nut, and a plastic thread cover. In use, the wing-nuts are just tight enough to keep the pivot box from rotating when it's not being pushed.

To move the binoculars in azimuth, the viewer moves from side to side and swings the whole arm around. The bearing for this movement (Figure 11.3, *overleaf*) is Ebony Star Formica riding on Teflon blocks attached to the tripod head. You can substitute any kind of stiff plastic surface for the Formica, and plastic feet used on the bottom of furniture legs for the Teflon, if needed. Drill the round base of the central pivot with the hole-saw after gluing the Formica. The $1\frac{1}{4}$ in (32 mm) diameter is flexible. Its only requirement is that it be larger than the $\frac{5}{8}$ in (16 mm) nut at the base of the tripod shaft.

We found the $\frac{5}{8}$ in × $\frac{3}{4}$ in (16 mm × 20 mm) bronze bushings in our local hardware warehouse store. They happen to have just the right inner diameter to slip easily over the tripod shaft. If you can't find similar bushings, you can drill slightly smaller holes and just have the wood moving next to the metal shaft. As noted on the diagram, you should attach the carriage bolts before attaching the sides. And don't forget that the exact $7\frac{1}{2}$ in (19 cm) spacing between the carriage bolts is important, as described above.

The weight box (Figure 11.4, *p. 186*) carries the counterweights which balance the binocular-pivot box-head assembly at the opposite end of the swing

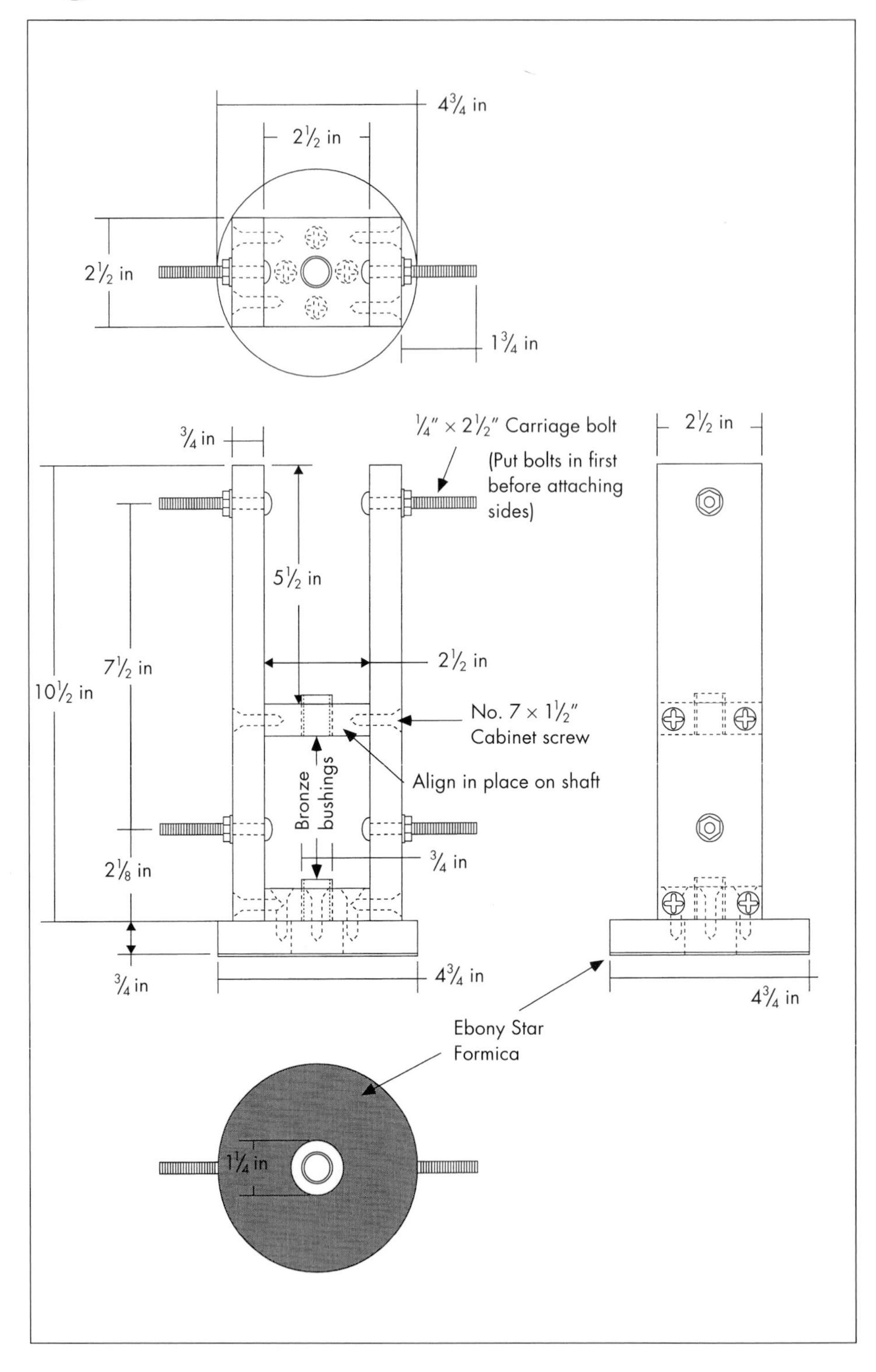

4¾ in
2½ in
2½ in
1¾ in
¾ in
¼″ × 2½″ Carriage bolt
(Put bolts in first before attaching sides)
2½ in
5½ in
2½ in
7½ in
10½ in
No. 7 × 1½″ Cabinet screw
Bronze bushings
Align in place on shaft
¾ in
2⅛ in
¾ in
4¾ in
4¾ in
Ebony Star Formica
1¼ in

Figure 11.3 The central pivot.

arm. The weights are the same 4 in (100 mm) diameter plates, used on dumbbells or barbells in weight-lifting, that are used as the pivot box counterweight. To balance my pair of 11 × 80 binoculars, which weigh 5 lb (2.25 kg), and the pivot box counterweight, the eight plates shown give 10 lb (4.5 kg) of counterweight. As shown in the picture, the weight box sits near the end of the swing arm. For lighter/heavier binoculars, you would adjust the balance by removing/adding plates in the weight box, and adjusting its position. The weights are held by $1\frac{1}{8}$ in (30 mm) dowel cut slightly under $1\frac{3}{4}$ in (45 mm) long. The wing-nuts never come completely off; they are merely tightened when you assemble the mount. At disassembly, they are loosened, and the weight box slides off as one unit with the weights held securely inside.

The shaft on the tripod head (Figure 11.5, *p. 188*) is a $\frac{5}{8}$ in × 8 in (16 mm × 200 mm) bolt with its head cut off and rounded with a file. Make sure in the store that the bronze bushings you have slip over the entire length of the bolt. In our store, I found that some of the bolts were very slightly too large near the head, which meant that the bushing would get stuck part way. The bushings are held in the wood by friction, and the fact that they had some oil on them which swelled the wood slightly over time. An alternative might be to roughen the edge of the bushing with a file and press-fit them. The bolt was threaded only about an inch, and it is these threads which hold the $\frac{5}{8}$ in (16 mm) nuts above and below the tripod head. The hinges were Stanley 6 in (15 cm) light-duty strap hinges, which meet in the middle and are held with a wing-nut, and 2 in (50 mm) utility hinges to join the legs to the head.

To set up the tripod (Figure 11.6, *p. 189*), you remove the wing-nut, which is on the permanently attached 1 in bolt on one of the strap hinges. Put the holes of the other two strap hinges over the nut and press down, so that the hinge holes snap in place on the nut. When the bolt is through the three hinge holes, put the wing-nut back on and tighten. The hinges act as spreaders and push the legs apart slightly. To break the tripod down, you simply remove the wing-nut, pop the hinges off the bolt and put the wing-nut back on, ready for the next time. You can leave the tripod feet as shown, or you can put on some rubber feet from crutches, or furniture legs. I attached a short length of nylon strap to one of the legs and installed a quick-release buckle which I purchased at a fabric store.

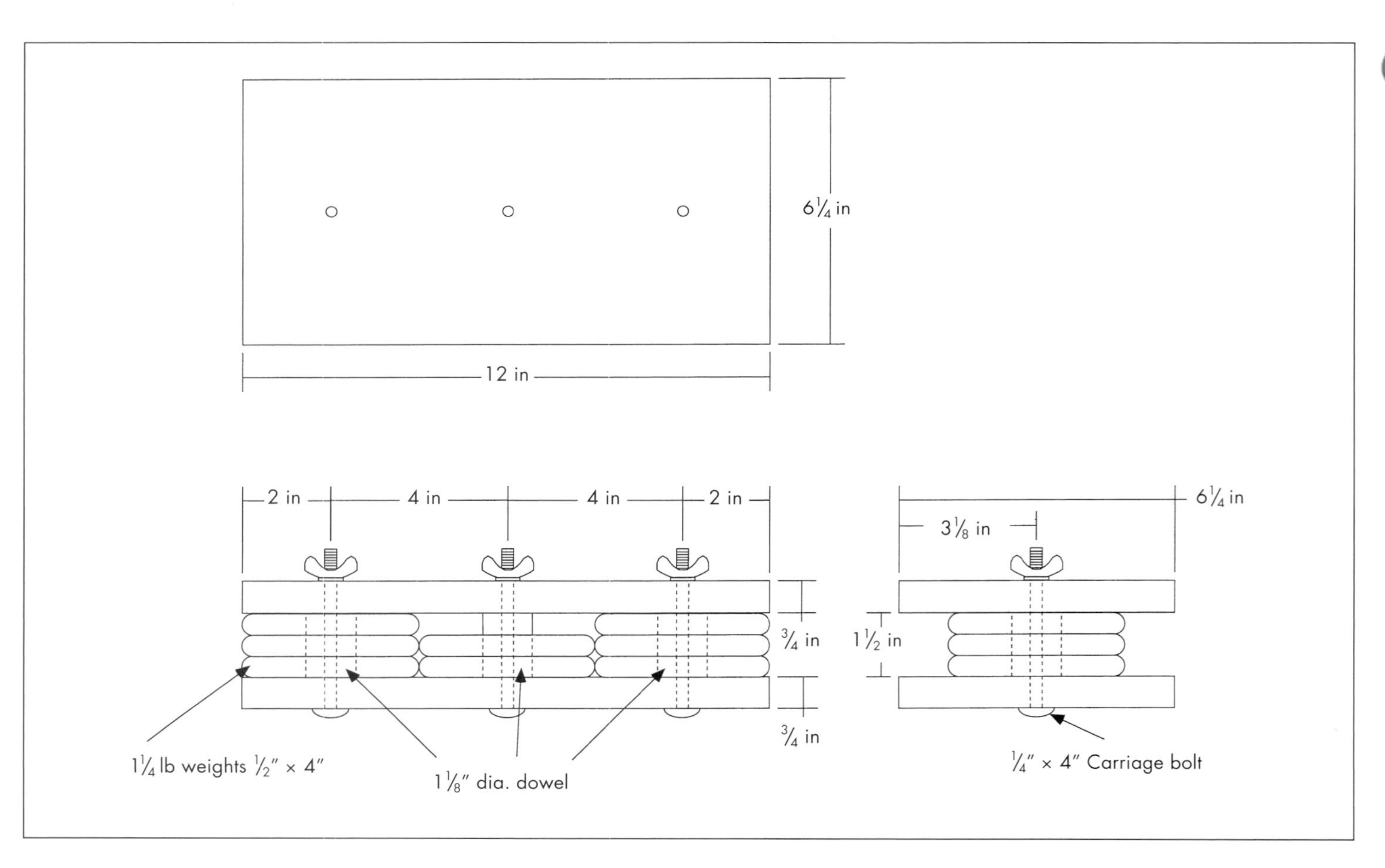
6¼ in
12 in
2 in
4 in
4 in
2 in
6¼ in
3⅛ in
¾ in
1½ in
¾ in
1¼ lb weights ½" × 4"
1⅛" dia. dowel
¼" × 4" Carriage bolt

Figure 11.4 The weight box.

When the legs collapse together, you snap the buckle together to bind them, and you can then carry the tripod by one of them.

The arms (Figure 11.7, *p. 190*) attach to the head assembly and central pivot. I used a nylon washer between the arm and the $\frac{1}{4}$ in (M6) nut on each of those parts in order to provide smooth movement. On the outside of each arm is another nylon washer, a $\frac{1}{4}$ in (M6) wing-nut and a white plastic thread cover.

To use this mount, you first open the tripod as described above. It will provide a solid base of support. You then take the main arm assembly, without the weight box or binocs, and put the central pivot on the tripod shaft. Next, you slide the weight box onto the end of the swing arm, holding it carefully until you tighten it down. You will do a final position adjustment later. Slowly lower the weight box and let go when it's resting on the tripod. The head assembly end will rise to its maximum height. Then, with your binocs in your right hand, pull the swing arm down until it's horizontal and, while holding it with your left arm, hold the binocs against your chest with your left hand and reach round with your right hand to attach them to the L-bracket. (Don't worry, you will probably find your own techniques and shortcuts to putting it together anyway.) Now that the binocs are attached, you should loosen the four wing-nuts on the central pivot and the four on the head assembly, so that the swing arm can move with minimum friction, and move the weight box to achieve final balance. Then tighten the wing-nuts and start having fun. To disassemble, simply reverse the process.

For convenience, I attached Velcro to the swing arm and Velcro to my lens caps, which makes them much easier to find in the dark. Depending on what kind of wood finish you use, you also might have to add some friction between the weight box and the end of the swing arm. A single layer of electrical tape on the swing arm, which contacts a single layer on the inside of the weight box, works well.

The completed binocular mount is shown in Figure 11.8 (*p. 191*). I hope you have as much fun building it as I had. I learned that the overall philosophy when building any kind of equipment is to make something that breaks down into a few easily transportable parts without having any free pieces which have to be kept track of, or could get lost. Ideally, you would also like to be able to set up the equipment with no tools, either.

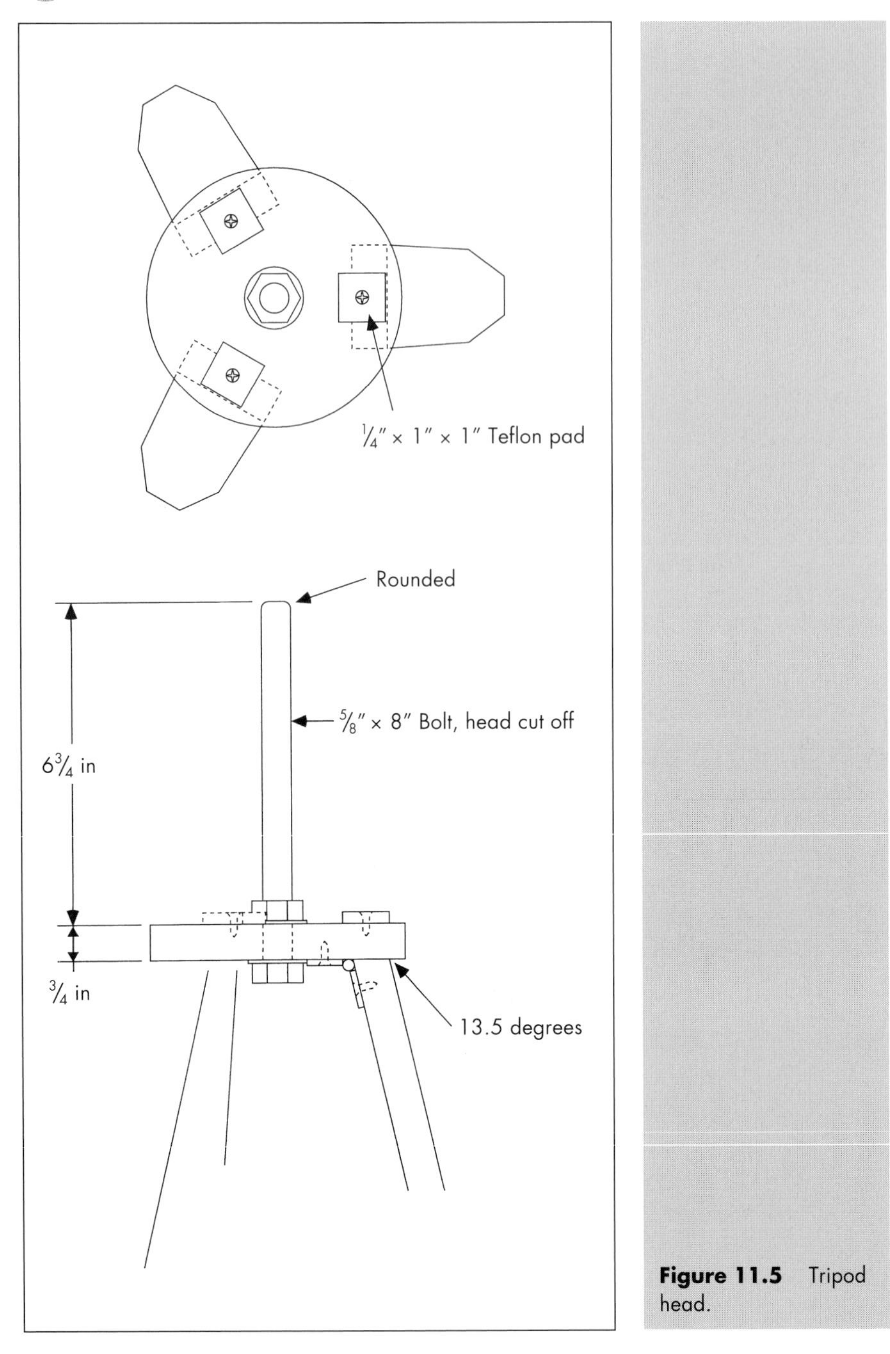

Figure 11.5 Tripod head.

The wing-nut on the tripod spreaders is the only free piece of hardware, and even if it gets lost, the tripod will still work. From the comments I have received at

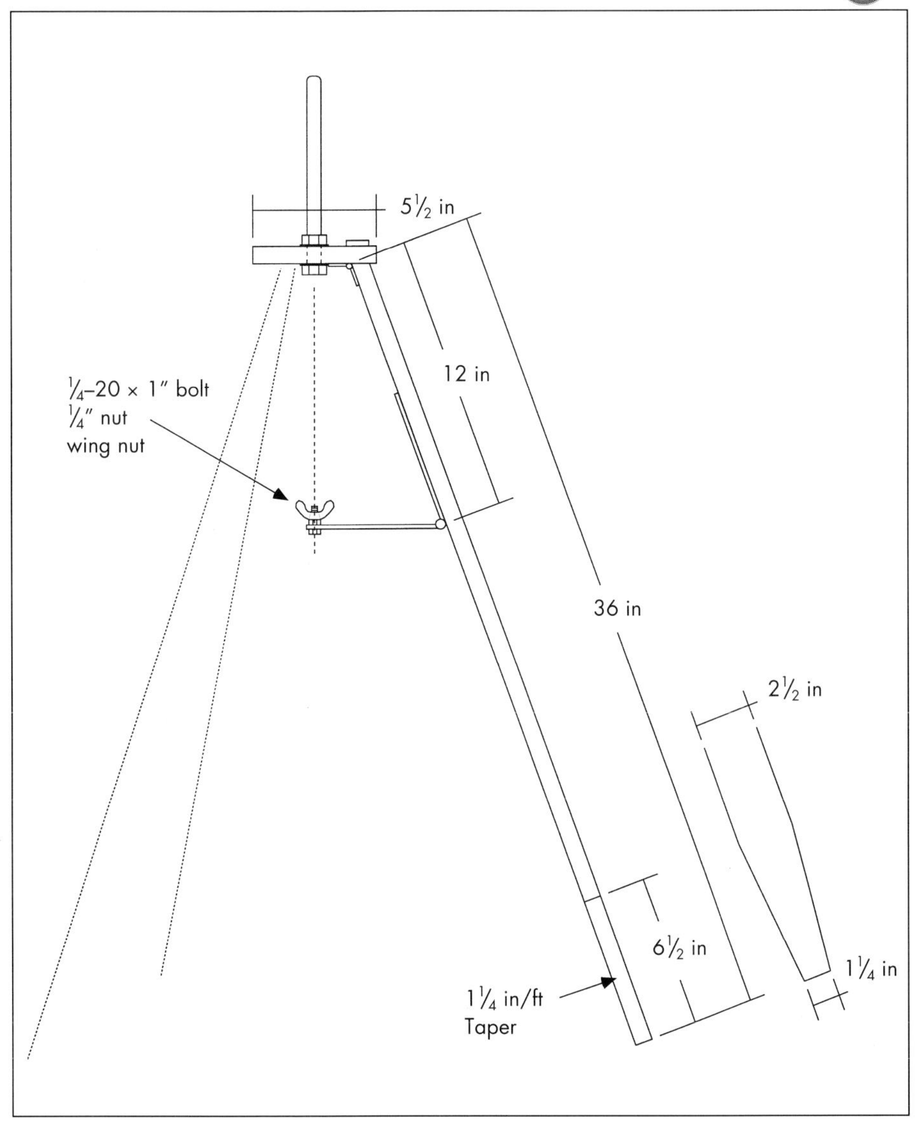

Figure 11.6 Tripod.

star parties and public observing events, as well as my own experience, I know how nice it is to simply point and look with both eyes. This mount makes the bread and butter of amateur astronomy – observing – as easy and stress free as possible.

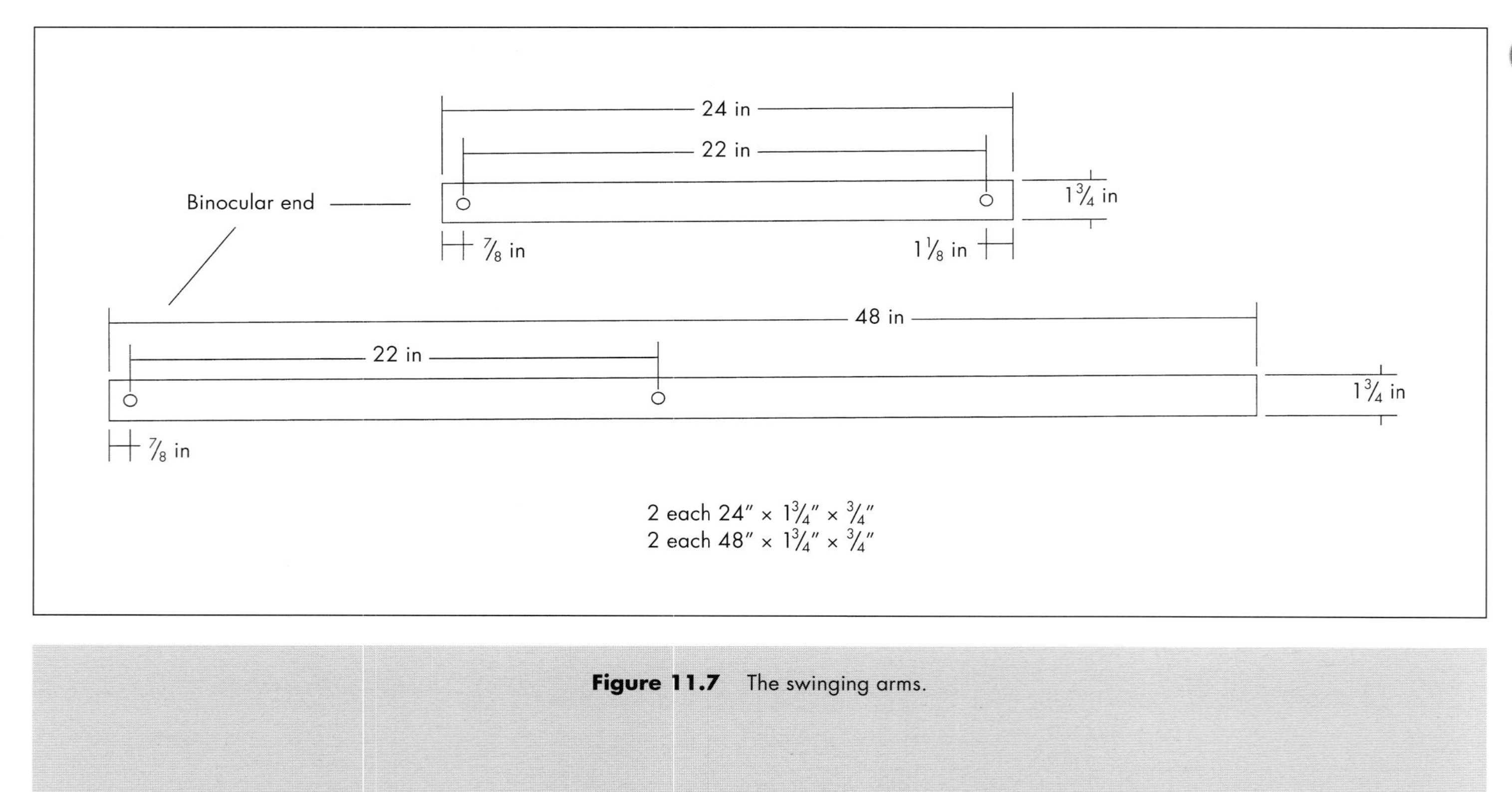

Figure 11.7 The swinging arms.

Figure 11.8 The completed binocular mount.

Part IV

Astrophotography

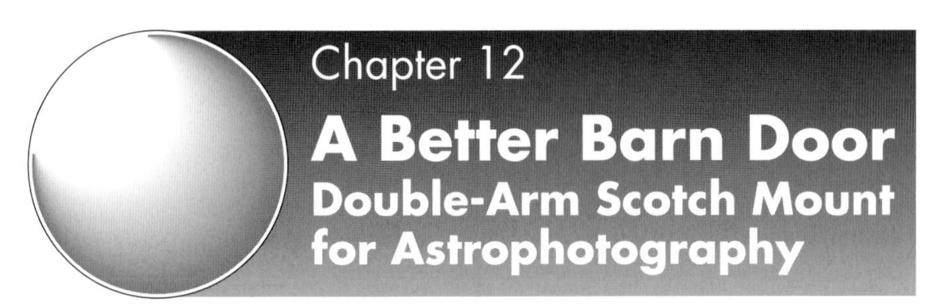

Chapter 12
A Better Barn Door
Double-Arm Scotch Mount for Astrophotography

Stephen Tonkin

Astrophotography with a 35 mm camera is usually done piggyback on an equatorially mounted telescope or on one of the several small equatorial mounts sold specifically for the purpose. The ATM who wishes to achieve the same ends through spending time rather than money has traditionally made a hand-driven "barn door drive". During the late 1980s, ATMs made several improvements to the original simple design. Excellent astrophotography is possible with this motor-driven camera mount, which can be built in a weekend. It offers more accurate tracking than conventional barn door designs, and it incorporates a stepper-motor controller which can be adapted to other applications.

The Scotch Mount

An inexpensive route to astrophotography has long been the Scotch mount, also known as the barn door drive (after its appearance) or the Haig mount (after its inventor). A double-arm mount offers much greater accuracy than conventional designs.

In its simplest form, the Scotch mount consists of two boards of wood, hinged at one end and driven near the other by a bolt which passes through the lower board and bears on the upper one (Figure 12.1, *overleaf*). If it is to be hand-driven, it is conveniently made so that one revolution of the bolt is made in one minute in order that the handle may be synchronised

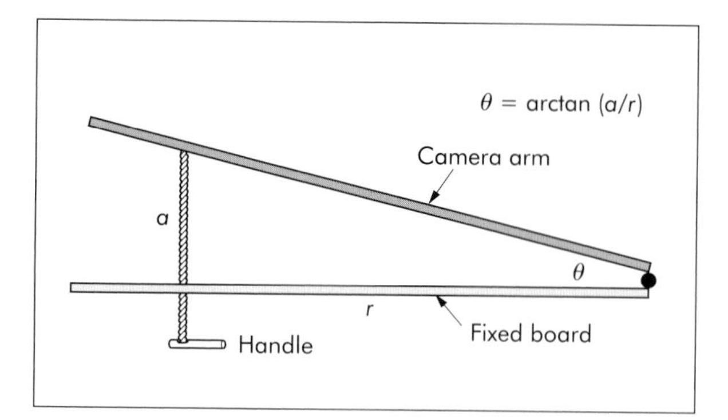

Figure 12.1 The right-angle or tangent drive.

with the second-hand of a watch. This is the *tangent drive* configuration. A tangent drive will be close to sidereal rate for only a very small angle (see Figure 12.4). If it is carefully made, it will track for 5 to 10 minutes before trailing becomes apparent with a 50 mm lens.

Improving the Basic Design

Accuracy of tracking can be improved by building the mount in an isosceles configuration (Figure 12.2). The drive is then across a cord of the arc through which the arm is driven and exposures of up to 20 minutes are possible with a 50 mm lens.

An isosceles drive is much more difficult to build, since both ends of the drive bolt must swivel about an axis parallel to the hinge. If this extra effort is to be taken, it makes sense to improve the tracking accuracy even further if this can be achieved simply.

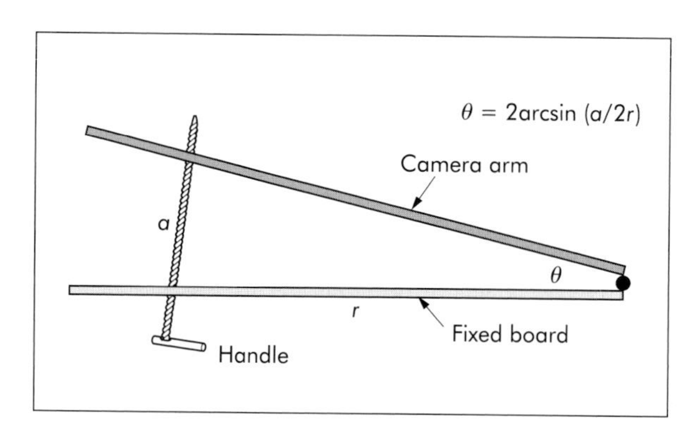

Figure 12.2 The isosceles drive.

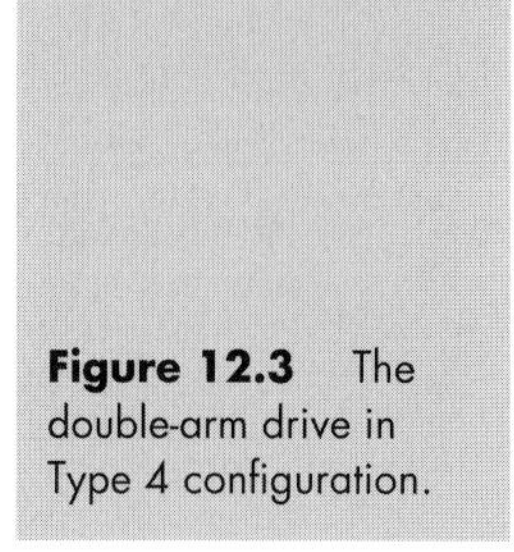

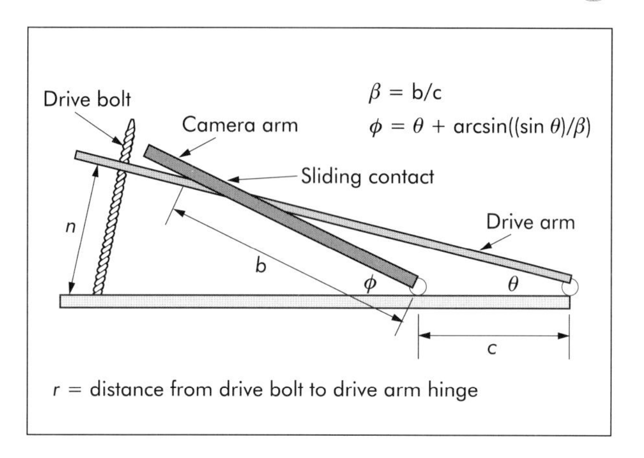

Figure 12.3 The double-arm drive in Type 4 configuration.

Many ways of improving the tracking accuracy have been tried, but the easiest to make with simple tools must be the double-arm configurations investigated by Dave Trott (Trott, 1988). The configuration with the greatest potential is Trott's Type 4 configuration (Figure 12.3), which can provide excellent tracking for nearly an hour, by which time errors resulting from atmospheric refraction would exceed the error of the drive (Figure 12.4).

Figure 12.4 Tracking errors for different Scotch mount configurations.

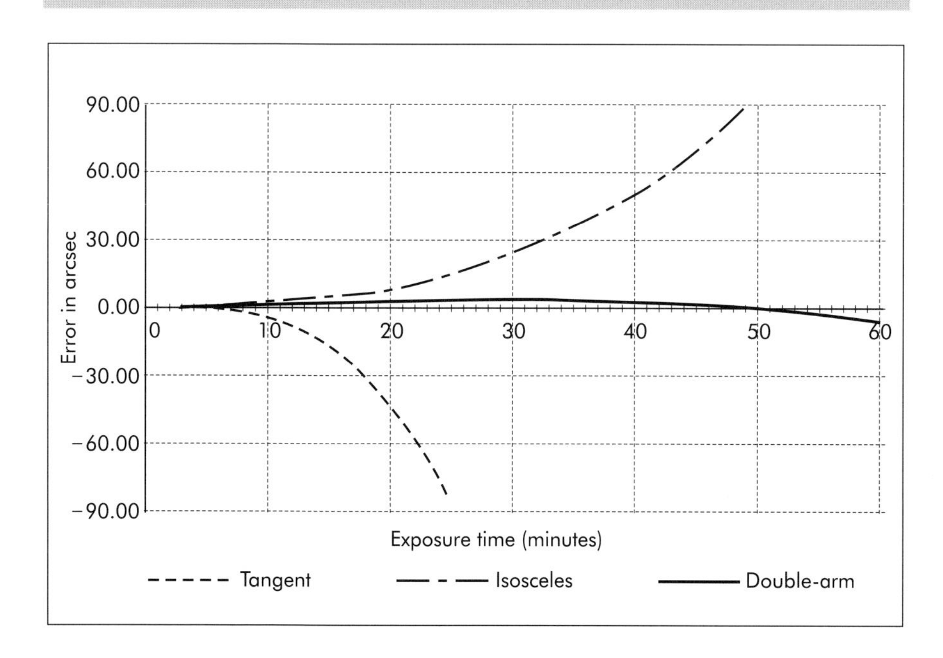

In the double-arm mount, the moving arm of the isosceles configuration drives a second arm upon which the camera is mounted. The duration of accurate tracking is determined by the parameter β, which is the ratio *b/c*. The angle ϕ through which the camera arm moves is given by

$$\phi = \theta + \arcsin(\sin\theta / \beta)$$

where θ is the angle through which the drive arm moves. Accurate solutions exist when β is in the region of 2.000 to 2.186 (Sinnott, 1989).

Design

The double-arm mount shown in Figures 12.9 *(p. 202)* and 12.11 *(p. 205)* uses a 6 mm (M6) threaded rod to drive it. I chose this size because its thread pitch of 1 mm results in a drive of moderate dimensions and has the secondary advantage that it simplifies calculations. Given a motor drive rotating at 1 rev/min, a solution of the equation above gives, to the nearest mm, the following dimensions:

$r = 333$ mm $\quad b = 254$ mm $\quad c = 116$ mm

However, since the accuracy of the drive depends upon the ratio *b/c*, it is advantageous to design the mount such that the length of *b* can be adjusted in the event that errors creep in during actual construction.

The accuracy of tracking will also depend upon an accurately driven motor. I decided to construct a simple crystal oscillator/driver for a stepper motor. RC oscillators are less expensive to make, but their accuracy suffers from the disadvantage of their being temperature-sensitive.

Construction

The mount shown in Figures 12.9 and 12.11 has boards of medium-density fibreboard (MDF), which is very simple to work using hand-tools (Figure 12.5). If it is properly sealed with varnish, it does not warp easily, and is dimensionally stable compared with most commonly available softwood timber. Constructors who have not previously worked with MDF should note that it is an advantage to countersink any pilot

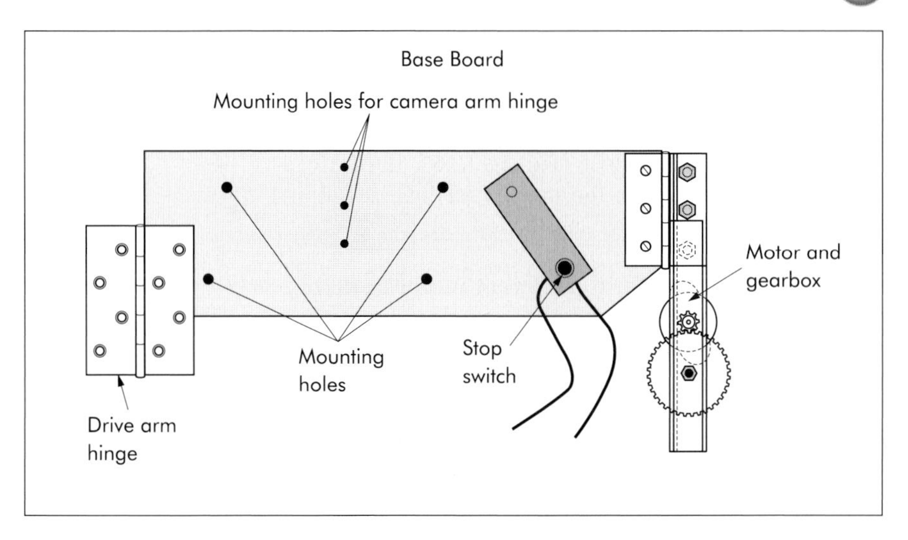

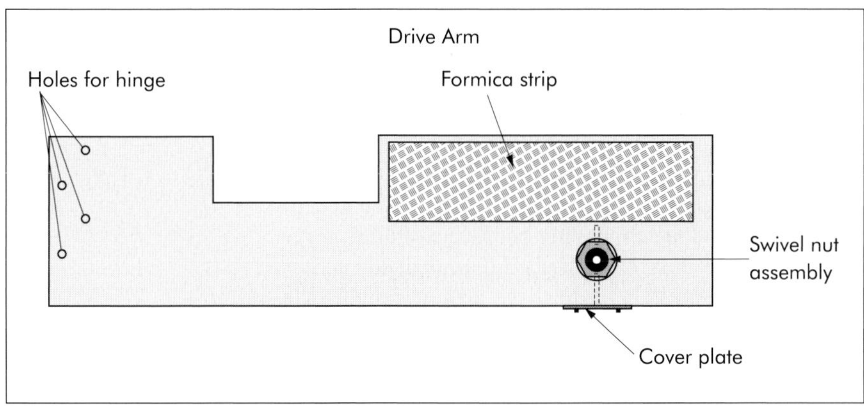

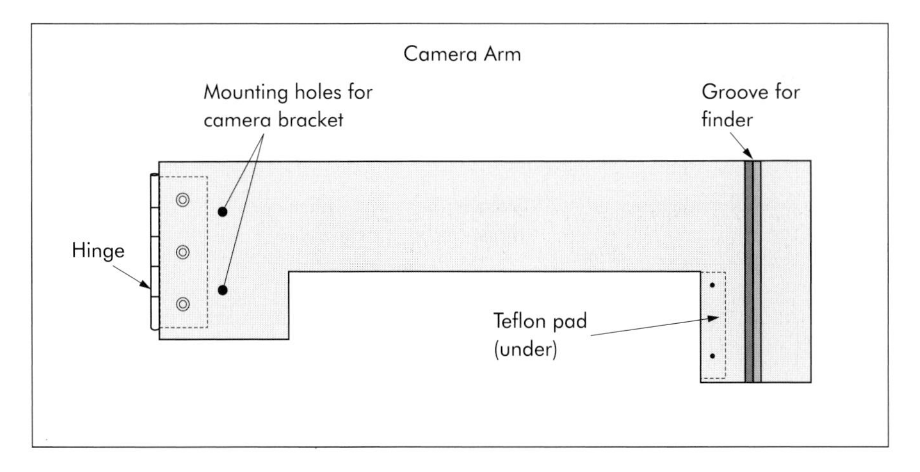

Figure 12.5 The three boards which comprise the double-arm drive.

holes before inserting screws, in order to prevent the material around the hole lifting above the surface of the board.

Aluminium angle was cut and drilled to form the motor-mount and gearbox (Figure 12.6), the latter being necessary because the motor, salvaged from a broken dot-matrix printer, has a step angle of 7.5°. A 5:1 gearbox permits the oscillator–driver circuit to drive the rod at 1 rev/min. Different motors will, of course, require different gearing, or even no gearing at all. Bushes for the threaded rod were made from cylindrical brass rod-connectors and were pressed, using a vice, into holes in the aluminium. The motor/gearbox assembly was then mounted on the fixed board using a stout brass hinge. All hinges used in this project must be solidly constructed with no play in the joint, which must move freely. Good quality brass hinges usually meet these requirements.

The drive-arm must be drilled to accept a nut assembly (Figure 12.7) which can rotate about an axis parallel to that of the hinges. I made this assembly by filing six equi-spaced grooves into the inside of a 12 mm (M12) nut and using a vice to press a 6 mm nut into the larger nut. I drilled opposite flats of the larger nut to accept 3 mm (M3) bolts which, having been decapitated, serve as swivels. A 2.5 mm hole is drilled diametrically across the larger (28 mm) hole in the drive-arm, parallel to the hinge axis, to accept the swivels.

You should insert the assembly into the drive-arm as follows: First, screw the upper swivel into its hole in the drive-arm. Then put the nut assembly into its hole and impale it upon the swivel. Lastly, insert the lower swivel and close the swivel hole with a metal plate. If

Figure 12.6 The motor and gearbox assembly.

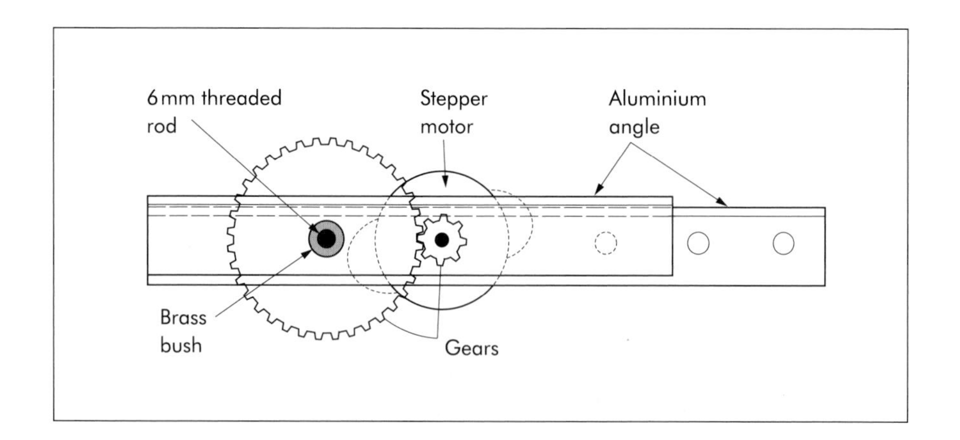

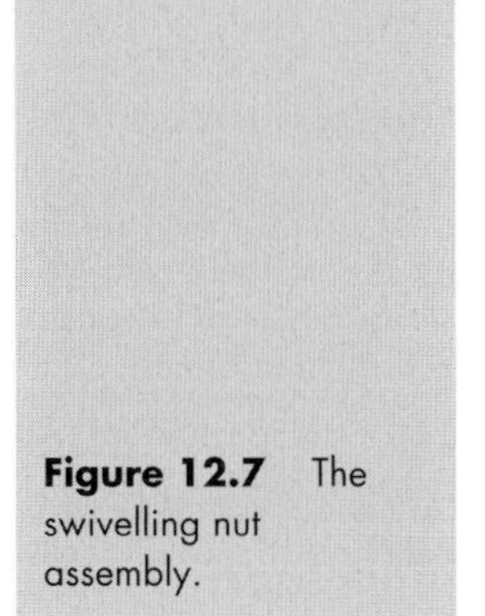

Figure 12.7 The swivelling nut assembly.

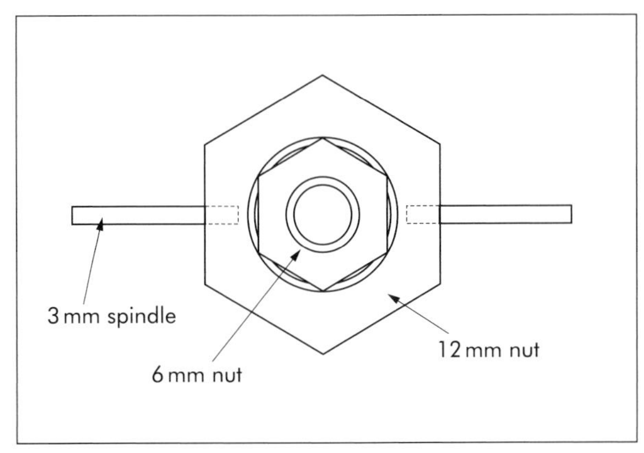

the swivels are a tight fit in the holes in the drive-arm, you can fine-adjust the position of the swivel assembly along its axis by rotating the swivels with a pair of long-nose pliers.

A strip of textured Formica (edging-strip for kitchen work-surfaces is suitable) should be glued along the drive-arm where the camera arm will bear upon it.

The completed drive-arm should then be mounted on the fixed board with another hinge, ensuring that the axis of this hinge is parallel to that of the one holding the motor/gearbox and that distance from the hinge axis to the centre of the drive-rod is accurate.

You next assemble the camera arm. Attach a ball-and-socket camera mount to a stout bracket made of about 40 cm aluminium strip approximately 20 mm by 3 mm (Figure 12.8, *overleaf*). This is screwed to the camera arm, which is in turn screwed to a third hinge, which is itself screwed to the fixed board, again ensuring that the hinge axis is parallel to those of the other hinges and that it is placed the correct distance, c, from the drive-arm hinge. A piece of Teflon must be attached to the camera arm where it bears upon the Formica on the drive-arm. It is as well to remeasure the distance between the drive-arm and camera arm hinges in order that you can position the Teflon so that the distance of the bearing surface from the camera arm hinge axis, b, gives a value for β which is as close as practicable to the optimum value.

You complete the mechanical part of the mount by screwing it to a frame, cut at the angle of your latitude, which attaches it to the tripod or whatever else will support your mount (Figure 12.9, *overleaf*). The frame is secured to the tripod with a central 10 mm bolt.

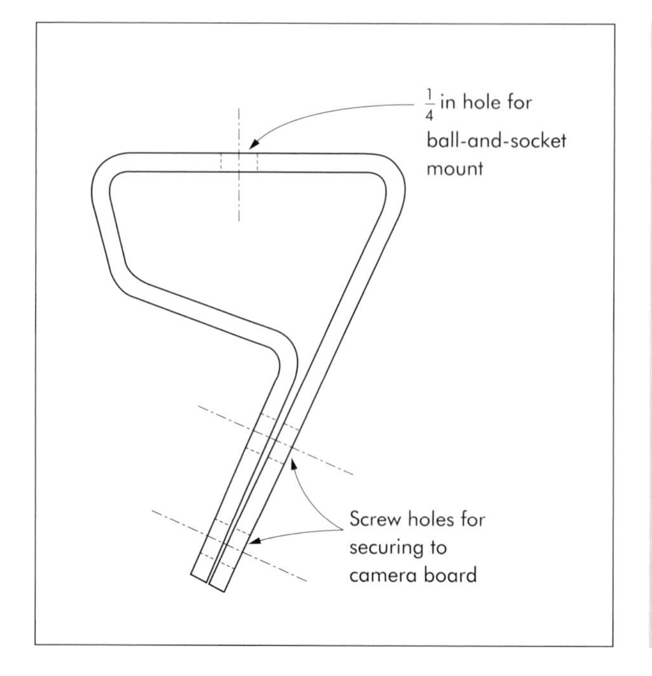

Figure 12.8 Camera bracket.

Figure 12.9 The mount on its tripod. The box-frame, through which it attaches to the tripod, also contains batteries, mains adaptor, and car cigarette lighter lead.

The Oscillator/Driver

The drive electronics (Figure 12.10) are based on a circuit published on the World Wide Web by Ray Grover (Grover, 1994). The output of a 4.194304 MHz

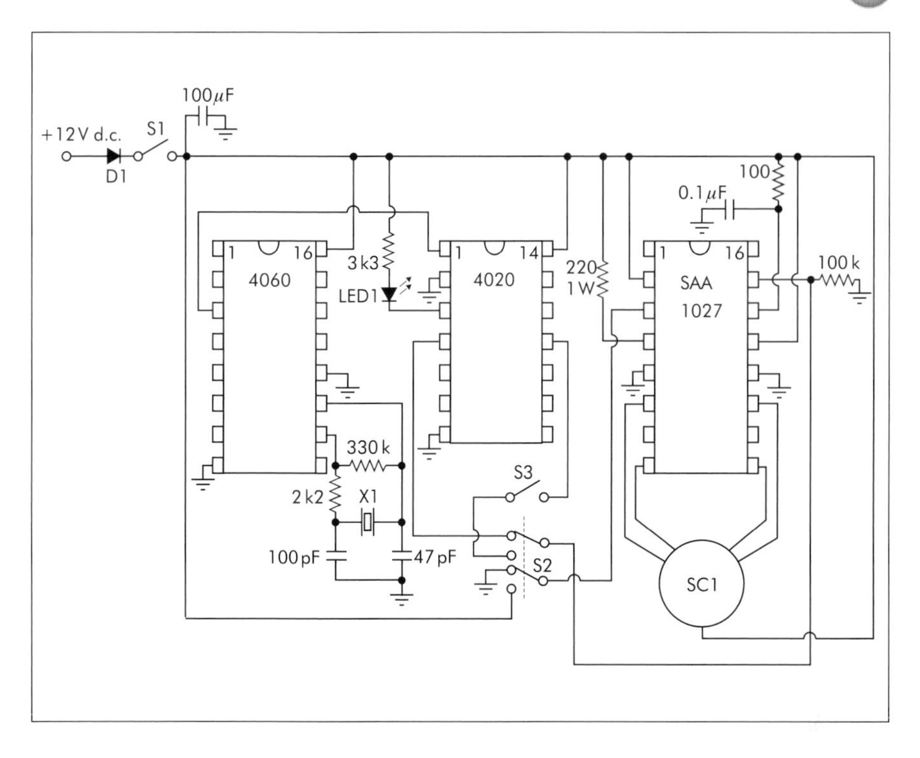

Figure 12.10 The motor control circuit.

crystal (X1) is divided by the CMOS chips, 4060 and 4024, to give a 4 Hz output to the SAA1027 stepper-motor driver. Each of the stepper motor (SC1) inputs is connected to an output of the SAA1027. If you are uncertain of the correct sequence of the stepper motor input leads, some experimentation will be necessary to establish this – this will not harm the motor. For more information on stepper motors, see the excellent resource, *Jones on Stepper Motors* (Jones, 1996).

Rewinding the drive-screw is effected by switch S2, which reverses the direction of the stepper motor by raising the voltage on pin 3 of the SAA1027. It also connects the 64 Hz output (pin 11) of the 4024 to the SAA1027 input, giving a rewind rate 16 the drive rate. S3, a normally closed pushbutton switch mounted on the fixed board, breaks the circuit when the boards come together. The LED flashes when the circuit is powered.

The components can be mounted on a piece of stripboard. The ICs are conveniently (and more safely!) mounted in sockets which are soldered to the stripboard. Although some electronics buffs are horrified by this approach, more flexibility in the use of the strip board is possible if you remove from the IC sockets before they are soldered to the board. The ICs are inserted into their sockets when soldering is complete. The circuit is housed in a project box with flying leads to the motor and switch S3.

The circuit draws about 0.25 A. It should not be connected to an unregulated supply in case the voltage surges above the safe limit of 15 V. The diode D1 protects the circuit against accidental reverse-polarity connection. A 12 V lead–acid battery is a convenient power source and, if the power lead is fitted with the appropriate plug, the circuit can be driven from a car's cigarette lighter socket. Alternatively, a multi-purpose regulated d.c. power supply, such as those available from electrical department stores, will allow you to power the mount from the mains.

Polar Alignment

There is little point in making an accurate mount if it cannot be accurately polar-aligned. The polar alignment is effected by a simple finder of "projected-pinhole" type, as described in Chapter 3. If the sky is viewed with both eyes open, one of them looking into the finder, the pinhole is projected onto the background stars. The finder is mounted on the camera arm, conveniently at the end most distant from the hinge (Figure 12.11). It is adjusted until there is no discernible movement of the pinhole against a distant object when the camera arm is rotated about its hinge – it is then parallel to the axis of the camera arm hinge and can be fixed in place. Silicone (RTV) adhesive is suitable for this purpose.

As the celestial pole is 44 arcmin from the star Polaris (α Ursae Minoris), the pinhole can be made larger so that its illuminated disc subtends double this angle, that is, 1.5°. The pinhole size can be achieved through trial and improvement – I initially achieved it accidentally and, fortunately, recognised its usefulness before I discarded the "ruined" pinhole. If you place the disc so that it is on the Kochab (β Ursae Minoris) side of Polaris, with Polaris on its rim, you can obtain a polar alignment to better than 10 arcmin. If you do this, it is essential to have a method of dimming the LED – I now use a 5 kW potentiometer, but dimming can just as easily be achieved if you cover the eye-end of the finder with a small piece of plastic lens cut from

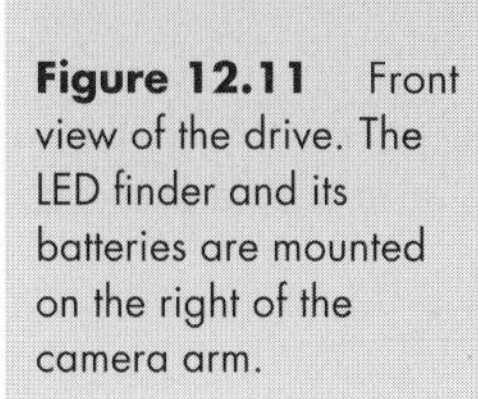

Figure 12.11 Front view of the drive. The LED finder and its batteries are mounted on the right of the camera arm.

discarded sunglasses. The error of polar alignment will most likely be the largest error affecting the mount. An alignment error of 10 arcmin will give a tracking error of about 2.5 arcsec per minute.

Performance

The mount performs very well indeed when properly polar-aligned (Figure 12.12, *overleaf*). Even 25-minute exposures with a 50 mm lens show no evidence of trailing when the star images are viewed under a microscope. I have not yet attempted longer exposures, owing to the light-polluted nature of the local skies.

While it does not afford the potential for guided photography, its performance matches, and possibly exceeds, that of some commercially available camera mounts intended for unguided photography. It does so for a very modest financial outlay at the cost of less than a weekend in the workshop, thereby putting the possibility of high-quality astrophotography within the reach of everyone.

Figure 12.12 Comet Hale–Bopp, 6 minutes, 135 mm lens, *f*/2.8.

Hardware Requirements

No attempt is made to give a minutely detailed parts list since, in a project like this, both design and construction depend upon the ingenuity of the constructor and the adaptability of available materials.

Approx. 1.5 m × 100 mm × 12 mm MDF, plywood or timber (for boards)
Timber for the rest of the mount
3 good-quality 75 mm hinges
Ball-and-socket mount for camera
Approx. 120 mm × 6 mm threaded rod
A stepper motor
Gears (if necessary, dependent on stepper motor)
Approx. 200 mm Formica edging strip
40 mm × 6 mm × 6 mm Teflon
Various nuts; bolts; screws; aluminium strip and angle; brass bushes; as required.

Calculation of Drive-Arm Length

If $b = 254$ mm and $\beta = 2.186$,

then c = 254/2.186 mm
= <u>116.2 mm</u>

ϕ should change at 2π radians per day = 360/1436°/min

Solving $\phi = \theta + \arcsin(\sin\theta/\beta)$, we find that the equation is satisfied when

$\theta = 0.172°$/min

Pitch p of 6 mm thread = 1 mm

Stepper motors are commonly available with step angles of 7.5° (48/min) and 1.8° (200/min):

Motor with 7.5° Step Angle and 4 Hz Input to SAA1027

Motor turns with a period of 360/(7.5 × 4) s

= 12s

Using this drive rate will produce a very long drive-arm, resulting in a heavy and unwieldy mount, hence gearing of 5:1 is required to attain a drive rate of 1 rev/min.

$\theta = 2 \arcsin(p/2r)$

where r is the length of the drive-arm

$r = 2p/2 \sin\theta$
= 1/sin (0.172°) mm
= <u>333.1 mm</u>

Motor with 1.8° Step Angle and 4 Hz Input to SAA 1027

Motor turns at a rate of 360/(1.8 × 4) s

= 50 s

This is conveniently close to 1 min, in which time the screw will advance $60p/50$ mm.

$\theta = 2 \arcsin[(60 \times p)/(50 \times 2r)]$,

where r is the length of the drive arm

$r = (2 \times 60 \times p)/(50 \times 2 \times \sin\theta)$
= 6/5 sin (0.172°) mm
= <u>399.5 mm</u>

Electronic Drive Components

ICs

CMOS 4060 divider/oscillator	1	
CMOS 4024 divider	1	
SAA1027 stepper motor controller	1	

Crystal

4.194 304 MHz quartz crystal	1	(X1)

Diodes

LED, red, miniature	1	(LED1)
1N4003 or similar (200 V, 1 A)	1	(D1)

Capacitors

100 μF, 16 V electrolytic	1	
0.1 μF ceramic disc	1	
47 pF monolithic ceramic	1	
100 pF monolithic ceramic	1	

Resistors

100 W, 0.6 W	1	
220 W, 1 W	1	
2 k2 W, 0.6 W	1	
3 k3 W, 0.6 W	1	
100 kW, 0.6 W	1	
330 kW, 0.6 W	1	

Switches

SPDT toggle, subminiature	1	(S1)
DPDT toggle, subminiature	1	(S2)
Push-to-break button	1	(S3)

Miscellaneous

Plug and socket for power connection; Vero stripboard; box to house circuit; 7/0.2 wire; DIL sockets for ICs; LED clip.

Signal Frequency on 4024 Pins

Pin	Frequency (Hz)
3	2
4	4
5	8
6	16
13	32
11	64
10	128

Chapter 13

"Skypod": A Simple Sky-Tracker for Astrophotography

E.G. Mason

Most designs for camera-tracking mounts relate to the northern hemisphere and require aligning on Polaris. New Zealander Euan Mason built this camera drive which can easily be adapted for use in either hemisphere. This project requires the careful work of a craftsman, if it is to be made well enough for the accurate tracking of which it is capable. It is highly suited to the ATM with advanced woodworking skills and access to a well-equipped workshop. As Dr Mason's photographs demonstrate, it is entirely worth the effort.

Introduction

If you live in the Southern Hemisphere, have you ever gazed mournfully at designs for barn door mounts, knowing full well that on most sites you would struggle for hours to get the hinge pointing at sigma Octans? If so, then have I got a device for you! You can align the tracking mechanism with the celestial pole in just a few minutes even during the daytime, and it requires no expensive tripod. It is easily portable, and costs more in construction time than in materials. I use it for astrophotography and as a mount for a 4.5 in Newtonian reflector. Ease of daytime polar alignment makes solar observing a snap. Come to think of it, there are advantages here for astronomers in the Northern Hemisphere as well!

With a good workshop and careful construction you can build a skypod to track for reasonably long periods and with camera lenses up to 200 mm. I have used mine to record the central starfields of our galaxy and also the Southern Cross with a single-lens reflex camera, a 50 mm lens, 400 ISO slide film, and tracking for 10 minutes and 5 minutes respectively (Figures 13.1 and 13.2). I have beautiful images of objects such as η Carinae, globular and open star clusters and comet Hale–Bopp taken with a 200 mm lens!

I am a self-taught amateur hewer of wood, with a workshop containing a bench saw and a mounted crosscut saw. I think these are the minimum bench machines needed to make a skypod accurately enough. If you are superbly skilled then you may be able to make the components with only hand-held tools, but I certainly needed the bench tools. I'm not a cabinet-maker, but neither am I a slug when it comes to wood-work. I made my first wooden aeroplane as a wee lad (do you ever wonder what the proverbial "first wood-working project" was before we invented aeroplanes?), and I've been working with wood off and on all my life. I'm not boasting here; it's just that I wouldn't like you to embark on a week of construction only to find your woodworking skills weren't quite up to the task. If you have patience and you can cut and fit wood so that it's perfectly square, you'll be able to handle this project.

Figure 13.1 The Sagittarius region taken using skypod, a 50 mm lens, 400 ISO film, and tracking for 10 minutes.

Figure 13.2 Crux and η Carinae taken with skypod, a 50 mm lens, 400 ISO film, and tracking for 5 minutes.

Before we begin discussing how to build a skypod, let me give credit where it is due. I didn't invent this device, and I don't know who did. I based my design on similar machines used by members of the Hamilton Astronomical Society in New Zealand. Like most useful gadgets, this one has elements that have been used before, and I don't claim to be the originator of those ideas either. My skypod has what I believe are enhancements over the gadgets I saw in Hamilton, and I offer you this chapter in the hope that you might further improve on the design when you build your own. Now that's out of the way, let's explore the features that make a skypod so useful.

The Basic Features of a Skypod

The completed skypod is shown in Figure 13.3 (*overleaf*). It consists of six essential features. After you understand their functions you'll be in a position to begin improving on the design described here.

Four of the features allow you to position the device with the right orientation and keep it there. These features are listed below:

Figure 13.3
Skypod assembled and ready for action.

- A sturdy tripod keeps the base of the machine still once you have it correctly oriented. The main leg of this tripod holds the moving parts which track the heavens.
- A plane table with a compass allows you to correct for magnetic declination and orient skypod's main leg along the north-south line.
- A pendulum with a large protractor orients the length of the main leg exactly perpendicular to the celestial pole.
- A simple tubular bubble level ensures that the front face (plane) of the main tripod leg is perpendicular to the celestial pole.

The other two features are attached to the main leg of the tripod:

- A flat arm beam rotates against the main tripod leg and a threaded bolt mechanism allows you to control the speed of rotation so that it exactly counters the Earth's rotation with respect to the stars. Given the right distance between the point of rotation and the place where the bolt touches the arm, you turn the bolt at a rate of one complete rotation per minute.

- A small altazimuth mount attached to the top of the arm beam is used to point your camera or telescope at the object you wish to track.

Construction

Read through this entire chapter before you purchase anything. This is not software; you *do* need to read the instructions, and there is no help button. You may decide to substitute other materials for some of the ones I used, and dimensions may change to suit those new materials. In particular, the aluminium box beam and the map compass may prove difficult to find. I've made suggestions where changes might be made.

In many cases the dimensions are not critical, and I admit that I made most of them up as I went along. The only absolutely critical measurements are those relating to the speed of rotation and angles associated with the mean latitude where you plan to use the machine. All other measurements can be changed a bit to suit materials at hand, so long as they all fit together and bearing surfaces are adequately stable.

Materials Required

550[1] × 50 × 40 mm (22 × 2 × $1\frac{1}{2}$ in) aluminium box beam (3 mm ($\frac{3}{8}$ in) thick aluminium)
60 mm × 100 mm ($2\frac{1}{4}$ in × 4 in) sheet of 2 mm ($\frac{1}{10}$ in) plate aluminium
240 mm × 20 mm ($9\frac{1}{2}$ in × $\frac{3}{4}$ in) sheet of 2 mm plate aluminium
200 mm × 40 mm (8 in × 1.6 in) sheet of 3 mm ($\frac{1}{8}$ in) plate aluminium
150 mm (6 in) of 7 mm ($\frac{1}{4}$ in UNC) threaded brass rod and 2 nuts to fit
250 mm (10 in) of 5 mm ($\frac{3}{16}$ in UNC) threaded brass rod
30 mm ($1\frac{1}{4}$ in) length of standard $\frac{1}{4}$ in UNC ($\frac{1}{4}$ in 20 tpi) threaded brass rod with wing-nut and two thin nuts
4 off 50 mm (2 in) 5 mm brass wood bolts

---

[1] Note that this value will critically depend on the radius calculations (below).

4 off brass wing-nuts to fit 5 mm threaded rods and bolts
6 off brass nuts to fit 5 mm threaded rods and bolts
7 off brass washers for 5 mm threaded rods and bolts
1 off 8 mm ($\frac{5}{16}$ in) brass bolt with two washers and a locknut
50 off 30 mm ($1\frac{1}{4}$ in) brass woodscrews
1 map compass (or similar, with accuracy better than 0.5°)
1 plastic bubble level replacement
Approximately 1 square metre (1.2 square yards) of 13 mm ($\frac{1}{2}$ in) thick marine plywood
1200 × 34 × 44 mm (48 × $1\frac{3}{8}$ × $1\frac{3}{4}$ in) wooden beam (naturally resistant to decay or treated)
2 off 1200 × 100 × 13 mm (48 × 4 × $\frac{1}{2}$ in) wooden planks
130 × 130 mm (5 × 5 in) section of Formica- or graphite-covered 3 mm thick fibreboard (MDF)
300 × 24 mm (12 × 1 in) section of Formica- or graphite-covered hardboard

You may be wondering why expensive brass and aluminium hardware is needed. The reason is that iron or steel components might make the compass inaccurate.

The Main Tripod Leg

Begin by making the main tripod leg which will support the tracking mechanism (Figure 13.4). I used a box beam made of aluminium. The beam is hollow, and made of 3 mm ($\frac{1}{8}$ in) thick metal. The exact dimensions are not critical, but it must be perfectly square and rigid for its entire length. The length required will depend on the number of threads per cm of your threaded rod, which in turn determines the length of your rotating arm. You will need at least 100 mm (4 in) more than the radius of rotation for the rotating arm (see radius calculations below).

I chose aluminium box beam because I was sure it would be perfectly square, it wouldn't deform with moisture, and it allowed the main arm to be dismantled into an aluminium section and a wooden extension. Earlier designs in Hamilton used one solid piece of wood for the entire main tripod leg. If you choose to use wood to hold the machinery, you should select a piece with no knots, made of a wood which is dimensionally stable and well-cured, and you must make it perfectly square.

Cut an angle on one end of the box beam, along the 50 mm (2 in) axis. This angle will be equivalent to the

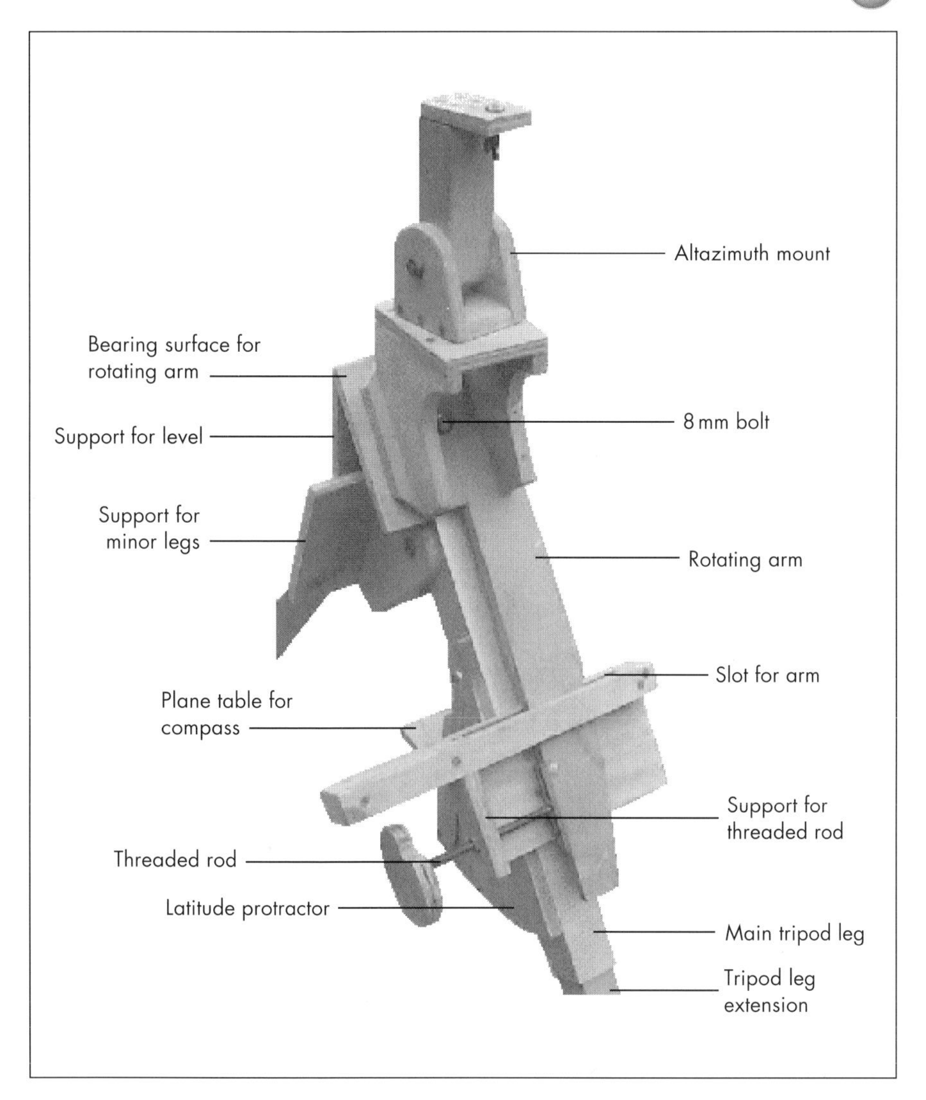

Figure 13.4 The parts of skypod which are described in the text.

mean latitude where you expect to use the device. In my case this was 38°, for Rotorua, New Zealand.

Four wooden plywood sections are rigidly attached to the main beam, and it is easier to attach them if you have pieces of wood inside the aluminium beam. I have a 200 × 34 × 44 mm (8 × $1\frac{3}{8}$ × $1\frac{3}{4}$ in) piece inside under the wooden latitude protractor and the threaded rod support, and a small 70 × 34 × 13 mm ($2\frac{3}{4}$ × $1\frac{3}{8}$ × $\frac{1}{2}$ in) piece of plywood underneath the support for the rotating arm. These provide good purchase for woodscrews, and the larger piece operates as a stop for the wooden extension to the main leg. Woodscrews should not be expected to cut into the aluminium, and you should drill holes of just the right size everywhere a brass screw or rod goes through aluminium.

Supports for the Rotating Arm and Level

A 140 × 140 mm ($5\frac{1}{2} \times 5\frac{1}{2}$ in) plywood section is attached to the top of the aluminium beam using four screws which go through drilled holes in the aluminium and into the wood inside. This piece acts as a support for the rotating arm. Sand the support's surface until it is as smooth and flat as possible. The four woodscrews holding this piece need to be recessed so that they don't interfere with the arm.

A 140 × 100 mm ($5\frac{1}{2} \times 4$ in) piece of plywood with an angled edge acts as a support for the level. Two small triangles placed between this and the 140 × 140 piece (above) allow it to be attached while still allowing room for a locknut on the 8 mm ($\frac{5}{16}$ in) bolt. Do not attach this piece until you have the rotating arm in place. I mention it now because you will probably want to cut this and the previous piece from the same 140 mm ($5\frac{1}{2}$ in) wide piece of plywood.

Latitude and Azimuth Protractors

Make two protractors from plywood. One is used with the pendulum to set the latitude and the other is a plane table for the compass which will allow you to set the deviation of the compass between the true pole and the magnetic pole. The angles required will depend on the latitudes where you want to use the gadget and the amount of magnetic deviation.

For the latitude protractor, make a pendulum either from aluminium or from a string and weight. I prefer an aluminium one because it is more robust and can swing without catching on the protractor (Figure 13.5).

Check on a local topographic map to see whether the magnetic deviation is west or east, as this will determine both the angular size of the plane table required and which side of the beam to attach it to.

Figure 13.5 The latitude protractor ensures that skypod is at the right azimuth.

Allow for any expected increase in the deviation (that is, 0.5° over 5 years). A small triangle supports the plane table underneath. You may want to hinge the plane table so you can get it flat at a variety of latitudes (I plan to do this with mine now that I've moved to Christchurch). Attach the plane table to the latitude gauge and then attach the latter to the side of the main tripod leg using two woodscrews which are placed through holes in the aluminium and screwed into the wood inside the beam.

Purchase a compass that is as accurate as possible, but which can be easily mounted on the plane table (Figure 13.6, *overleaf*). I was fortunate enough to obtain an old "map compass" which has a long, 13 cm (5 in) needle but a narrow case. You need high accuracy from your compass, but over only a narrow range.

Figure 13.6 The compass plane table protractor ensures that skypod is in line with the earth's axis of rotation.

You may have to adapt the aluminium plate upon which the compass is fixed to accommodate your compass. The plate is fixed with a bolt and wing-nut to the origin of the protractor, and it has an indicator pointing to the magnetic declination scale.

Join the two protractors at an angle, so that when the device is set up the compass is level. When they have been joined, fasten them to the appropriate side of the main leg box beam by screwing through drilled holes into the wood inside the beam.

The Support for the Threaded Tracking Rod

The fourth piece to be permanently attached to the main leg is the support for the threaded tracking rod. Do not attach this piece until you have assembled all of it except the wood that covers the rotating arm. Do not drill the hole for the threaded rod until this piece is attached; that way you can make sure the hole is exactly the right distance from the pivot of the rotating arm.

The threaded rod support comprises a slot to keep the rotating arm straight when pressure is applied to it, and a small piece which will have a nut on each side to hold the threaded rod. Glue a strip of 3 mm ($\frac{1}{8}$ in) fibreboard along the bottom of the slot so that the rotating arm will slide neatly in the slot and always at the same distance from the main leg. Two small sections of plywood on either side of the slot ensure that the rotating arm fits snugly.

Before you attach this support and slot to the main leg you should calculate the radius of rotation required so that your skypod tracks the stars as the earth spins.

Radius Calculations

I find it convenient to turn the threaded rod once per minute after I've opened the camera shutter. Configuring skypod to this rotational velocity is simply a matter of having the correct distance between the pin holding the rotating arm and the point of contact between it and the threaded rod.

The first step is to count the number of threads on a long section of the threaded rod, and convert this value to threads/cm. Count the number of threads several times to make sure you have not made an error. If your section is X cm long, you simply take the number of threads and divide it by X. Do not round this value.

I'll briefly explain the calculations required. If you wish you can skip this bit and just use the formula below. The period of rotation of the Earth's axis with respect to the stars is 23 hours, 56 minutes and 4.099 seconds (1436.0683 minutes), slightly longer than the sidereal day due to precession of the equinoxes. The angular rotation required for one minute is therefore 360/1436.0683 or 0.250 684 45°. The length of arm required if you wish to rotate the rod once a minute is calculated from 2 times the tangent of half this angle (0.004 375 276). The formula is therefore:

$$L = 10 \div (0.004\,375\,276T)$$

where L is the length of rotating arm from the centre of the pivot to the threaded rod in mm, and T is the number of threads/cm of threaded rod. For my machine the value of L is 324 mm, but yours will depend on the number of threads/cm. The difference between your value and mine should be either added or subtracted from the length of the aluminium box beam for the main tripod leg.

The Rotating Arm and Mount Support

The rotating arm has a 130 mm (5 in) square section with rounded corners at the top where it fits onto the top of the main tripod leg, and a longer piece that has a face in line with the centre-line of this square section. Glue a layer of fibreboard with either graphite or Formica to the underside of the 130 × 130 mm portion in order to reduce friction as it turns. Drill an 8 mm ($\frac{5}{16}$ in) hole through the centre of the 130 mm square portion. I rounded off the three exposed corners of the 130 mm square portion, but this is not crucial.

Make a support for the altazimuth mount angled so that the top of the support will be level at your latitude. I cut circular edges to the side supports to allow easy access to the wing-nut holding the altazimuth mount, but if you use a ball joint instead of the altazimuth mount these cut-out edges are not necessary. Screw this support to the rotating arm with screws recessed so they will not catch as the arm rotates.

Assembly of the Rotating Arm and Main Leg Beam

Drill an 8 mm ($\frac{5}{16}$ in) hole through the centre of the support for the rotating arm and right through the aluminium and wood beneath.

Attach the support for the threaded rod to the main leg beam so that the rod can be placed with room to spare for a turning knob and perpendicular to the rotating arm at the point where it will push the arm (L mm from the centre of the 8 mm ($\frac{5}{16}$ in) hole).

Use the 8 mm ($\frac{5}{16}$ in) bolt with a washer on each side and the locknut to attach the rotating arm. Tighten it to the point where you can move the arm smoothly but also so that it would not move with the weight of your camera pulling it on one side.

Now attach a strip of plywood across the slot at the other end of the rotating arm, and screw the 140 × 100 mm ($5\frac{1}{2}$ × 4 in) level support in place.

The Threaded Rod and Push Point

Make a round, 8 cm (3 in) diameter turning knob (with a slot on the edge for your finger) out of plywood and attach it to one end of the threaded rod. Drill a small depression into the other end of the threaded rod where it will push against the rotating arm.

Drill a hole in the support for the rod so that the rod will push the rotating arm at distance L from its pivot. Use a small chisel or sharpened screwdriver to make some shallow hexagonal holes on each side of the tracking rod's hole. These hexagons should accommodate the nuts that go with the rod. With the nuts in place screw the rod through them so that they are tight on both sides of the rod's support. You may have to try several positions for one of the nuts before they are snug on each side. If you are lucky, you may find a T-nut that will do the job and that isn't made with iron, but I couldn't find one.

You need a sharp point at distance L on the rotating arm for the threaded rod to push on. The depression in the end of the threaded rod allows this point to seat neatly with little friction. My point is adjustable, but a small brass nail with its head removed, protruding from the right point, will suffice.

Tripod Legs and Level

Make a wooden extension to the main aluminium tripod leg so that it can fit neatly into the end of the main leg assembly. This fit must be very snug, but not so tight that the two pieces are inseparable.

Make a support for the other two minor legs of the tripod which consists of a 200 × 180 mm (8 × 7 in) rectangle of plywood with two triangular pieces separated by a piece of 40 mm ($1\frac{1}{2}$ in) wood. Round off the tips of the triangles that might otherwise butt up against other pieces of the beam. I cut a 40 mm ($1\frac{1}{2}$ in) slot in the top of the 200 × 180 mm piece so that the legs could be pulled further out. If you make higher triangles then this slot will not be necessary. Drill a hole right through the aluminium beam to accommodate the section of 5 mm ($\frac{3}{16}$ in) threaded rod which will hold the attachment pivot. Use a locknut and wing-nut with the rod to hold the pivot in place. The two remaining legs are held to the pivot with bolts and wing-nuts.

The exact dimensions and lengths of the tripod legs are left to you. Angles between legs of about 60° will make the tripod nice and stable, and a higher tripod will make it easier to point the camera but will also make the assembly more exposed to wind. My mount is about 1.7 m (67 in) high when it is set up.

Before attaching the plastic level you need to set the tripod up so that the pendulum indicates that the main leg is at the right angle for your latitude. Next place a carpenter's level across the top of the bearing surface for the rotating arm, make it level, and then tighten the tripod legs and their pivot support. You are now ready to attach the plastic level with the bubble it its centre (Figure 13.7). I made a small aluminium structure to

Figure 13.7 The level is the device which finally makes skypod point to the celestial pole.

hold the plastic bubble, but this may not be necessary if you buy one with holes already drilled. Purchase an adjustable one if you can.

The Altazimuth Mount

I made a little altazimuth mount to hold my camera, and the design is easy to copy from the photos. You may prefer, however, to use one of the ball-jointed mounts that camera shops sell. If you make your own, you need to include a $\frac{1}{4}$ in 20 tpi ($\frac{1}{4}$ in UNC) threaded rod with a locked wing-nut to thread into the base of the camera. When my camera is not mounted, I use another nut to hold the rod in place. If you use a ball-jointed mount, you may have to remove it until you've got the skypod aligned, otherwise it could affect the compass.

Using a Skypod

Figure 13.8 shows the skypod dismantled. I take about 5 minutes to set up mine before I open the shutter for the first photo.

Choose a spot away from large metal objects like cars or wire fences. These objects *will* affect your compass, and you'll be wasting your time and film if the skypod is poorly aligned.

To set a skypod up, first make sure that the correct declination is set on your plane table, and then orient

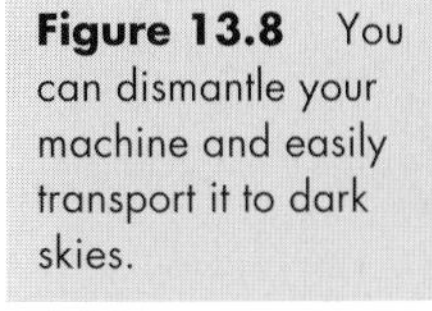
Figure 13.8 You can dismantle your machine and easily transport it to dark skies.

the tripod directly south (or north in the Northern Hemisphere).

Next adjust the two back legs so that the main leg is at the right angle for your latitude, measured by the pendulum and protractor. Recheck the alignment using the compass.

Then rotate the main leg assembly (the minor leg attachments allow this rotation) until the bubble indicates that it is level. Recheck all three measurements several times to make sure. Tighten the two wing-nuts holding the minor legs.

Attach your camera to the altazimuth mount. It's a good idea to have a shutter extension cable so that you don't deflect the camera when you open the shutter. Set the camera to "B". Using the altazimuth mount, point it at whatever you wish to record. Open the shutter and turn the knob on the threaded rod once a minute until the exposure is finished. I turn my knob 1/6 of a revolution every 10 s when I'm using a lens of 100 mm or less, and I try to keep the knob turning continuously for longer lenses. A friend made me a small box of electronics which beeps and lights up a light-emitting diode every 10 s. I like this timer because it's easily detectable in the dark without imposing any extra light. You may prefer to light up a watch or a small clock with a red torch or LED powered by some batteries.

If I Were Building Another Skypod ...

There are one or two things I would do differently if I were beginning again. I would make the plane table adjustable so that it was level at a variety of latitudes, and I would allow for better access to the locknut on the 8 mm bolt which holds the rotating arm in place. I'll leave these improvements up to you.

I wish you luck with your project, and many clear dark skies when it is finished.

Chapter 14 Building and Using a *Cookbook* CCD Camera

Al Kelly

Not long after the CCD camera became available to amateurs, some ATMs worked upon making this desirable device amenable to home construction. In 1994, *The CCD Camera Cookbook* was published, since when several hundred ATMs have made their own cameras. This is an attractive option, since the "Cookbook" CCD can be built for about half the cost of a commercial instrument. Although he had no knowledge of electronics, Al Kelly followed the instructions and built this device with which he now takes excellent images. He describes the construction process from a beginner's perspective.

The Concept

I stand as proof that anyone with sufficient motivation can build an excellent CCD camera. If detailed knowledge and experience with electronic components were required, my focuser would still only hold eyepieces. It continues to amaze me how successful Richard Berry, Veikko Kanto, and John Munger have been in bringing the intricacies of astronomical CCD camera construction into people's homes.

Since their *CCD Camera Cookbook* was first published in 1994, several thousand copies have been sold and it has been estimated that 4000 *Cookbook* cameras have been built. The *Cookbook* is aptly named. It is a recipe book for successfully making all the electrical and mechanical subsystems for a cooled astronomical CCD camera. It contains explicit instructions for consolidating these subsystems into a unit which is easily controlled by a low-end home desktop PC or laptop. All the necessary camera-control software is included, along with excellent programs for testing the electronic subassemblies as the project progresses.

While it is true that electrical engineers and electronic technicians may progress more smoothly than those of us without their particular expertise, it is also true that they will tend to experiment a bit more and often create more problems than they solve! All it really takes is desire for a CCD camera, the ability to read and follow directions accurately, about US$500, and some spare time. I purchased the *Cookbook* in the late spring of 1994 and completed the basic camera, fully tested and functional, in August 1994 and I have any number of friends who can vouch for my lack of abilities!

In the late 1980s and early 1990s I had many opportunities to use other CCD cameras, including the Photometrics Star I and the SBIG ST-4 and ST-6 cameras. These experiences convinced me that the direction I wanted to take with my hobby was CCD imaging. Unfortunately, the cost for the type of camera that would satisfy me was more than I was willing to pay, but all this changed with the publication of the Cookbook. The TC-245 chip is large enough for general-purpose imaging and the liquid cooling provides a very stable, low-noise receiver.

What skills and tools do you really have to have to build a CB245? Since pre-machined camera bodies and pre-printed circuit boards can be purchased (and usually are), a builder really only needs a soldering iron, a digital multimeter, and normal hand tools. Builders with machining skills and access to a metal lathe can save a few dollars by creating their own camera bodies from scrap aluminium. I am fortunate enough to have a friend who is an excellent machinist, so I had able assistance with the camera body. I built the rest of the camera on my kitchen countertop, following the words of the *Cookbook* as exactly as I could. My CB245 has been very durable and when I "fried" two of its components in 1996 through my own stupidity (I plugged two wires in wrong) I was able to troubleshoot it and fix it myself within 2 days!

Deciding to Do It

The promise of effective CCD imaging of deep-sky objects with exposures of only a few minutes has almost a magical allure. Coupled with the potential of being able to afford a high-quality CCD system, the enticement begins to bear a remarkable resemblance to the common astronomical malady of aperture fever (from which I suffered greatly several years ago). I bought the *CCD Camera Cookbook* knowing that I would learn from it even if I never built a camera. I had read only the first few pages when I knew that I had to give it a try. The descriptions were so straightforward and the systems so comprehensible that the investment in time and money seemed almost secondary to the siren call of the project. Also, the *Cookbook* had the necessary assurances that I need not understand everything about what I was doing to actually do it right. The best thing of all is that those assurances turned out to be absolutely valid!

Buying and Assembling Materials

Parts for the *Cookbook* camera system (Figure 14.1) are easy to obtain. The parts that are truly beyond the capability of most people to fabricate are available from a variety of sources, and University Optics (Ann Arbor, Michigan) sells complete parts kits. The *Cookbook* publisher, Willmann-Bell, sells pre-printed circuit boards for the major electronic subassemblies (the preamplifier card and the interface box). These make the wiring and soldering of the electronic components relatively easy for even a neophyte.

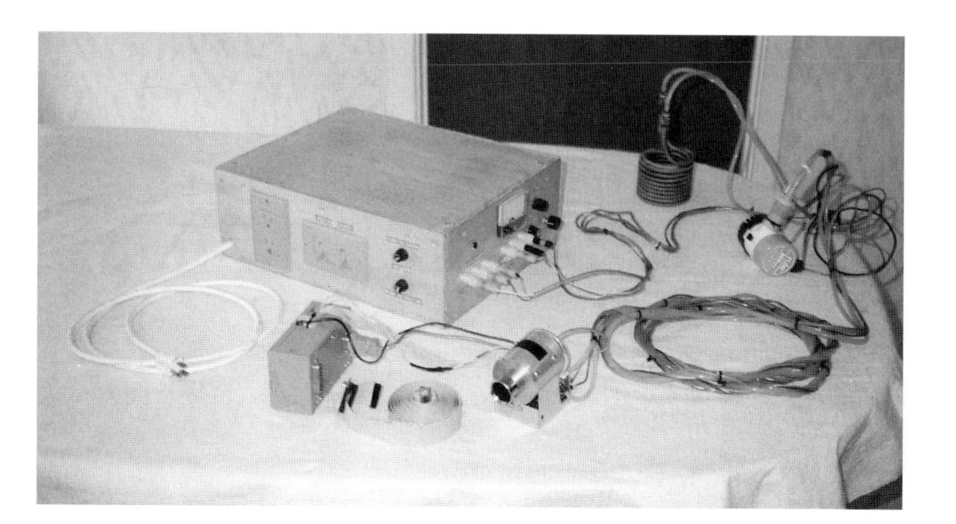

Figure 14.1 The different units of the *Cookbook* CCD system.

I ordered the pre-printed circuit boards from Willmann-Bell and several of the major electronic and hardware components from University Optics. While I was waiting for these to arrive, I decided to save a little money by shopping for most of the hardware and electronics parts (resistors, transistors, diodes, transformers, wire, hoses, copper tubing, and so on) from local (Houston, Texas) and mail-order sources. I expected this to be an exercise in frugality and learning about electronics parts. It was, but it was more an exercise in learning retail business patterns. I spent many hours sifting through surplus parts bins and studying stock catalogues. The amount of money I saved was probably less than the minimum wage for the hours I spent, but it did help the days pass while I waited for the mail-order parts to arrive.

Since the *Cookbook* construction process consists of the fabrication and testing of a series of subassemblies, ending with the final amalgamation and testing of the whole system, materials can be purchased and assembled in blocks. This allows work to be done while other parts are still on order or yet to be purchased from the local electronics outlet.

Construction and Testing

Following the *Cookbook* to the letter, the first subsystem I built was the power supply (Figure 14.2). This was a well-prescribed sequence, since the power supply consists of larger, less complex hardware, electrical, and electronic parts than most of the rest of the project. Inexperienced builders thus learn to read wiring diagrams and gain soldering experience with the easiest part of the project. Complete power supplies for the *Cookbook* system can be purchased, but I recommend the success of building the power supply as the proper start for any *Cookbook* builder.

The next subsystem is the interface box, which uses one of the pre-printed circuit boards and requires relatively detailed wiring and soldering of small electronic components. This was not a particularly difficult job, but I was introduced to more intricate and compact fabrication requirements and had to integrate parts into a housing ... excellent experience for the latter stages of the project.

Figure 14.2 The power supply unit.

Purchase of the *Cookbook* includes the software required to run the camera from a personal computer and software programs for testing the subassemblies as they are fabricated. Starting with the interface box, I had to manufacture simple hardware to connect the subassemblies to the parallel port of the computer. This hardware allowed the completed subassemblies to be tested with the provided software. I was able to uncover any problems before they were difficult to pinpoint and correct. All I had to be able to do was follow instructions and read a simple multimeter. The *Cookbook* even described how to do that.

The more difficult subsystems to fabricate, the preamplifier card and the camera head itself, were done next. After successful completion and testing of the power supply and the interface box, these more demanding parts became intriguing and very enjoyable to complete. Besides, the light at the end of the tunnel was quite visible! The *Cookbook* instructions and test programs were extremely valuable here, and these intricate efforts proceeded quite rapidly.

The final steps were construction of the liquid cooling system, assembly of the camera head, and complete operational testing. The cooling system was relatively simple and easy. The fabrication of the camera head, which included wiring and mounting of the CCD chip itself, mating the hydraulic cooling chamber, sealing the optical window, and sealing several wiring penetrations of the camera body, was probably the most demanding part of the project. However, since this step was the culmination of a very rewarding personal effort, it went quickly and happily. I found that any problems uncovered during final operational testing could be diagnosed rapidly, since all subsystems had already passed many tests. The first images produced by your own CCD camera are incredibly satisfying and become irreplaceable keepsakes!

Using the *Cookbook* Camera

Once the completed system is powered up, cooled down (which takes about 15 minutes), and the camera head is ensconced in the focuser of your favourite telescope, it is time to turn on the computer and run the software. The image acquisition software provided with the *Cookbook* works very effectively and can function on any PC. It does require about 590 KB of the 640 KB of RAM available in lower PC memory to run properly, so the operator will want to ensure that any unnecessary memory-resident programs are removed. The program runs out of DOS and can be run out of a DOS shell from Windows if sufficient RAM is kept clear.

The software communicates with the camera via the parallel port. The way I view it is that the computer thinks it is talking to a printer which can create data and upload binary files! Use of the speedy parallel port interface is one of the beauties of the *Cookbook* system. Unlike many commercial cameras which use a serial interface and require the better part of a minute to download and display a completed image, acquired *Cookbook* images download and display in just a few seconds. As long as we are using modern technology capable of providing nearly instant gratification, we may as well use it fully. It is amazing how long 30 seconds can seem to be while one is sitting in the dark in front of a computer!

The acquisition software can receive image files from the camera, display them, save them, subtract dark frames, and show a histogram of the image data. It can be used in quick-response modes to rapidly display images for focusing and for finding objects. It can be used to integrate one image at a time by keystroke or in an automatic image integration mode to take high numbers of images without operator intervention. It is a fully capable software which provides many other functions ... specifically, all functions necessary to acquire and store images. Two of the images produced by the author's camera are shown in Figures 14.3 and 14.4.

Figure 14.3 Globular Cluster M22, imaged by the author.

Figure 14.4 Stefan's Quintet, imaged by the author.

Modifications and Additions

I have departed from the published *Cookbook* design only in using a small submersible bilge pump rather than the recommended automotive windshield washer pump as the heart of the cooling system. The bilge pump is quieter and runs smoothly and dependably on its own regulated d.c. power supply. The windshield washer pump worked fine but kept me awake during imaging sessions!

After the builder has used the camera for a few months and has completely "burned in" the system, it is time to think about performing the one major modification suggested by the authors and published by Willmann-Bell. This is called the low dark current or LDC modification. For a few dollars, upgraded acquisition and testing software and fabrication instructions are provided. This upgrade is well worth the effort, since it greatly reduces the thermal noise generated by the camera, allowing much longer, higher-quality single exposures to be made. Of course, since CCDs are inherently linear data collectors, an unmodified camera can be used to acquire and stack (average) numerous short images, creating effectively much longer exposures; but the images of an upgraded *Cookbook* camera are much less noisy per unit time and make it much easier to get those "deep" images of faint objects. The electronic parts for the upgrade cost only a few dollars, and the modification can be accomplished and tested in one afternoon. For most people, it is not a good idea to try to incorporate the modification in the initial construction of the camera, since it is important to test and use the basic system for a while before changing it.

In order to do colour imaging, or any other filtered imaging, it is very useful to add a filter slide or filter wheel to the front of the camera. My engineering and machining genius friend, Andy Saulietis, designed a filter wheel for me which clamps onto the front of my camera in a very low-profile way (Figure 14.5). This is important, since the TC-245 chip is about 23 mm inside the camera's front surface and it must reach the focal plane of the telescope without producing vignetted images (losing light by all of the chip not being able to "see" the full diameter of the optical system). Reaching the focal plane can be difficult with some of the modern,

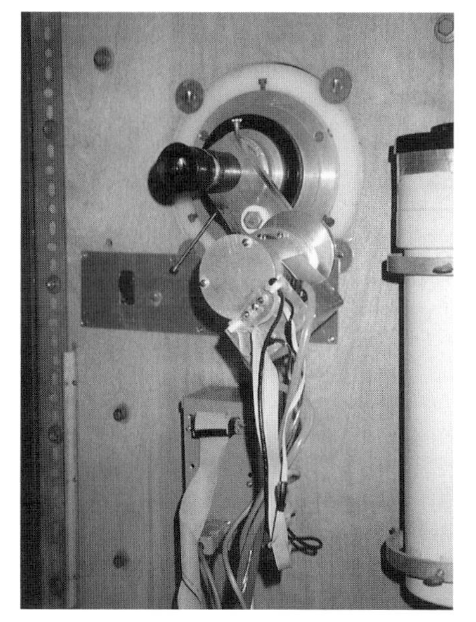

Figure 14.5 The eyepiece–CCD–filter assembly allows easier focusing and filter-changing.

fast optical systems, such as f/4 Newtonian telescopes. For three-colour imaging using red, green and blue filters, having a filter slide or wheel is important, since the filters can be introduced into the light path without having to remove the camera from the focuser. Removing and reinserting the camera head usually requires refocusing and engenders field rotation among the images which must later be reversed before the separate colour images are composited. For colour balance, an infra-red blocking filter must also be used, but this filter can be screwed into the front of the camera's slide tube and left in place for use with the other filters.

Image Calibration and Processing

Careful focus and tracking can provide data which results in excellent *Cookbook* images, but the data must be properly processed for the images to be seen in their best light. Proper processing means highlighting the signal from the object of interest by removing undesired signal and improving the overall signal-to-noise ratio (SNR). Images are considered to be properly calibrated when undesired signal from the camera's thermal and electronic sources has been removed and any distortions related to imperfections in the optical system have been corrected. This is commonly accomplished with the subtraction of a "dark frame" from the object image and division of a flat-field image into the object image. A dark frame is an image of just thermal and electronic signal taken by integrating an image with the camera closed off from any light source. A flat-field is an image of the imperfections in the optical system taken with the optical system evenly illuminated by a twilight sky or an artificial light source. I have found that although dark frames are almost always necessary for good *Cookbook* images, flat-field frames can often be ignored if the image is to be used only for "pretty picture" or sky search purposes. If these are the primary goals, then keeping the optical system (especially the camera's optical window) clean and well baffled, and assuring that the CCD chip does not produce vignetted images, will allow the user to frequently forego flat-fielding. Improving the SNR with the *Cookbook* system is best accomplished by stacking (averaging) several images to create an effectively much longer exposure.

What software is required to properly process *Cookbook* images? The answer is that any image-processing program which reads native *Cookbook* file formats, provides calibration functions, supports multiple image registration and averaging, includes scaling and filtering algorithms, and allows image files to be saved to commonly readable formats is sufficient. This sounds like a lot of capability, but it is available from many software sources. Richard Berry, well known as the former editor of *Astronomy Magazine*, author of many articles and books, image-processing software guru, and co-author of the *Cookbook*, has written a suite of software programs specifically for the *Cookbook* cameras which accomplish all of the above and a lot more. Each of these programs is inexpensive, and a *Cookbook* user can select only those programs needed for his or her individual needs. I use all of Richard's programs and find them to be very powerful and perfectly tailored to *Cookbook* needs. They are described in detail and can be ordered from Richard through his Internet web site at World Wide Web address (URL) <http://wvi.com/~rberry> I highly recommend them.

Several other excellent image-processing programs, such as Bruce Johnson's Megafix, Christian Buil's WinMIPS and Michael Newberry's MIRA, can calibrate and process *Cookbook* images. I have little personal experience with these, but each program has many proponents. As with all software, people tend to like what they have learned to use.

After images have been calibrated, averaged, scaled, filtered, or otherwise processed to the photographer's satisfaction, they should be saved to a format which is readable by more general graphics viewing and post-processing softwares, such as Adobe PhotoShop or Paint Shop Pro. Berry's software can save processed images in the TIFF format, which is universally readable. Monochrome or colour TIFF images can be brightened, darkened, or otherwise manipulated in post-processing and then saved in file-compression formats, such as JPEG or GIF, so that they can be efficiently sent to friends or published on Internet web sites for general enjoyment.

My Recommended Techniques and Recent Activities

In 1996 I finally reached my goal of producing tricolour CCD images with the *Cookbook* camera. Since then, I have been refining my techniques and recently have been involved in promulgating a new colour compositing process which combines data from highly detailed monochrome images with the colour balance and saturation from RGB tricolour images. I refer to this technique as MRGB processing, short for monochrome, red, green, and blue. Readers should go to my Internet web site at <http://www.ghg.net/akelly/> for further explanation and background on the MRGB revolution. The images included in this chapter have all been produced with the MRGB technique.

Following are short descriptions of the steps I take to produce an MRGB image. I hope that readers can gain some assistance from hearing details of one person's process:

1. Once the camera is cooled and carefully focused (with no filters in place), I find the object, centre it and take a sample integration, usually about 30 s long. Viewing the resulting image and its histogram, I look for areas of image saturation (stars with "blooming" streaks, and so on). If the image has very low SNR for

the object of interest, and there are no areas of significant pixel saturation, I will take longer integrations (but never longer than about two minutes to assure excellent unguided tracking). If there are areas of significant saturation and the SNR is good I will take shorter exposures to eliminate the saturation.

2. I shoot unfiltered images first, then install the IR filter and swing in the red filter for refocusing prior to taking the RGB images. With the filters in place and refocusing accomplished, I usually shoot red images first, green next, and blue last, just because that is the order of the filters on my wheel. Depending on the surface brightness and apparent SNR of the object, I may shoot only a few exposures of a few seconds each; or I may shoot dozens of exposures unfiltered and in each colour, creating stacked integrations of many minutes each. As a general rule for my particular CCD chip, I try to shoot at least twice as long in green and three times as long in blue as I do in red. My total unfiltered image exposure is usually at least as long as the red exposure duration, since I want the unfiltered monochrome image to have excellent SNR characteristics.

3. Sometime during the imaging session I will shoot dark frames totalling many minutes so that a high-quality master dark frame can be created later for image calibration. If it appears necessary, because of the optical system being used, that flat fielding should be performed during image calibration to eliminate optical defects, I will take from four to eight flat-field images through each filter and appropriate dark frames with which to calibrate the flats. With my Newtonians I almost never take flat-field images. I make them only if I am worried about photometric results, which is rare for me.

4. Back home, usually the next day, I load all the image files from the laptop to my desktop computer for faster processing. I will first create the master dark frame by averaging the numerous dark frame integrations in Berry's CB245 or MULTI245. Next, if necessary, I will create the master flat-field frames using the same averaging technique.

5. The next step is to calibrate, stretch, and average the images into M, R, G, & B stacks. I am careful to use only Berry's linear stretching transfer function in MULTI245 at this point, since I do not want to change the natural colour balance within and among the images. I am also careful not to stretch the images

Figure 14.6 Spiral galaxy NGC1232, from a tricolour image made by the author.

enough to create significant pixel saturation in terms of the 12-bit dynamic range of my system. These become my master calibrated stacks which are to be retained for all future processing.

6. Once the stacks are created I load them into Berry's QCOLOR software for compositing. In this software, the images are resampled to make the pixels square, rotated (if necessary) so that they are oriented exactly the same, scaled using the same low- and high-range values to assure a proper colour balance, and composited with sub-pixel registration. Also in QCOLOR, I will process the unfiltered M stack further with non-linear transfer functions or unsharp masks to bring out image intensity details. When the MRGB composite is correct, it is saved in a TIFF format for post-processing and final image production in Paint Shop Pro or Adobe Photoshop.

7. The final step is to use post-processing software (usually Adobe PhotoShop or Paint Shop Pro) to adjust the brightness and contrast of the image and to save it in 24-bit JPEG format.

One of the products of the method outlined above is shown in Figure 14.6. To create astronomical CCD images is to enjoy life!

Chapter 15 A Telescope Controller for Synchronous Motors

David Johnson

Reliable equipment has always been one of the goals of the ATM. There are some advantages to using synchronous motors on a telescope drive; not least, they are vibration-free and will therefore not resonate with the mount or telescope. David Johnson made this drive-corrector for use with the synchronous motors of his telescope drive. Ten years later it is still working perfectly, enabling him to take excellent CCD images from his home in the north of England.

Introduction

This unit was originally built as a low-cost means of taking celestial photographs. It is still cheaper to build a similar unit, rather than buying a commercial one. Anyone with a little experience in electronic construction should be able to complete this project. No special components are used; those listed in the parts list should be widely available from mail-order suppliers. Furthermore, no test equipment is needed, although an ordinary multimeter and a simple frequency counter may prove useful.

The design provides a safe means of powering mains-driven telescope mounts from a 12 V supply. The supply requirement is from 11 to 16 V d.c. at 500 mA. A sealed lead–acid gel cell, rated at 6 Ah, has sufficient capacity to run the unit for several hours.

This controller produces a variable frequency output for a telescope's right ascension (RA) drive, together with a switched output for declination. Varying the drive rate allows tracking corrections to be made during long-exposure photography. By omitting the hand controller (and relay board) it can be simplified for ordinary visual use.

Not all telescope mounts are suitable for use with this controller. In particular, drives using stepper motors cannot be driven with this equipment.

Circuit Description

This unit provides 230 V a.c. at 50 Hz for a right ascension (RA) drive and a switched, reversible-phase, 230 V a.c. supply for a declination motor. Both these outputs are isolated from ground and from the 12 V supply. The design is easily modified to 110 V 60 Hz output or for 12 V d.c. declination motors (Figure 15.1).

The 12 V input is fed via a 2 A fuse to switch S1 and series diode D1. IC5 is a 5 V regulator feeding IC1 to IC4, including the remote controller.

IC1 forms a free-running oscillator with a frequency determined by the value of resistance between pins 6 and 7. The oscillator has a nominal frequency of 10 kHz; this relatively high frequency is chosen so that a high-stability capacitor can be used for C10. This gives a minimum frequency drift with changes in temperature. Virtually identical oscillators are fitted to both the main and remote units. References to IC1 also include IC1A of the remote unit.

Resistors R4 and R5 are only fitted to the remote controller. Pressing PB1 (normally open contacts) places R4 in parallel with R3A; this raises the oscillator's frequency. Pressing PB2 (wired for normally closed contacts) adds R5 in series with R3A; this lowers the frequency. Switch S2 is fitted to the front panel of the main unit and selects either local or remote oscillators. The remote unit is connected via a latching DIN plug and socket, or similar connectors. The remote unit is only required for photographic tracking and can be omitted altogether for visual work.

IC2 and IC3 are wired to divide by 10, giving approximately 1 kHz and 100 Hz outputs. The output of IC3 is fed to the input of a flip-flop (IC4). This feeds the signal alternately to the bases of Tr1 and Tr2; these switch the

Figure 15.1 Controller circuit diagram.

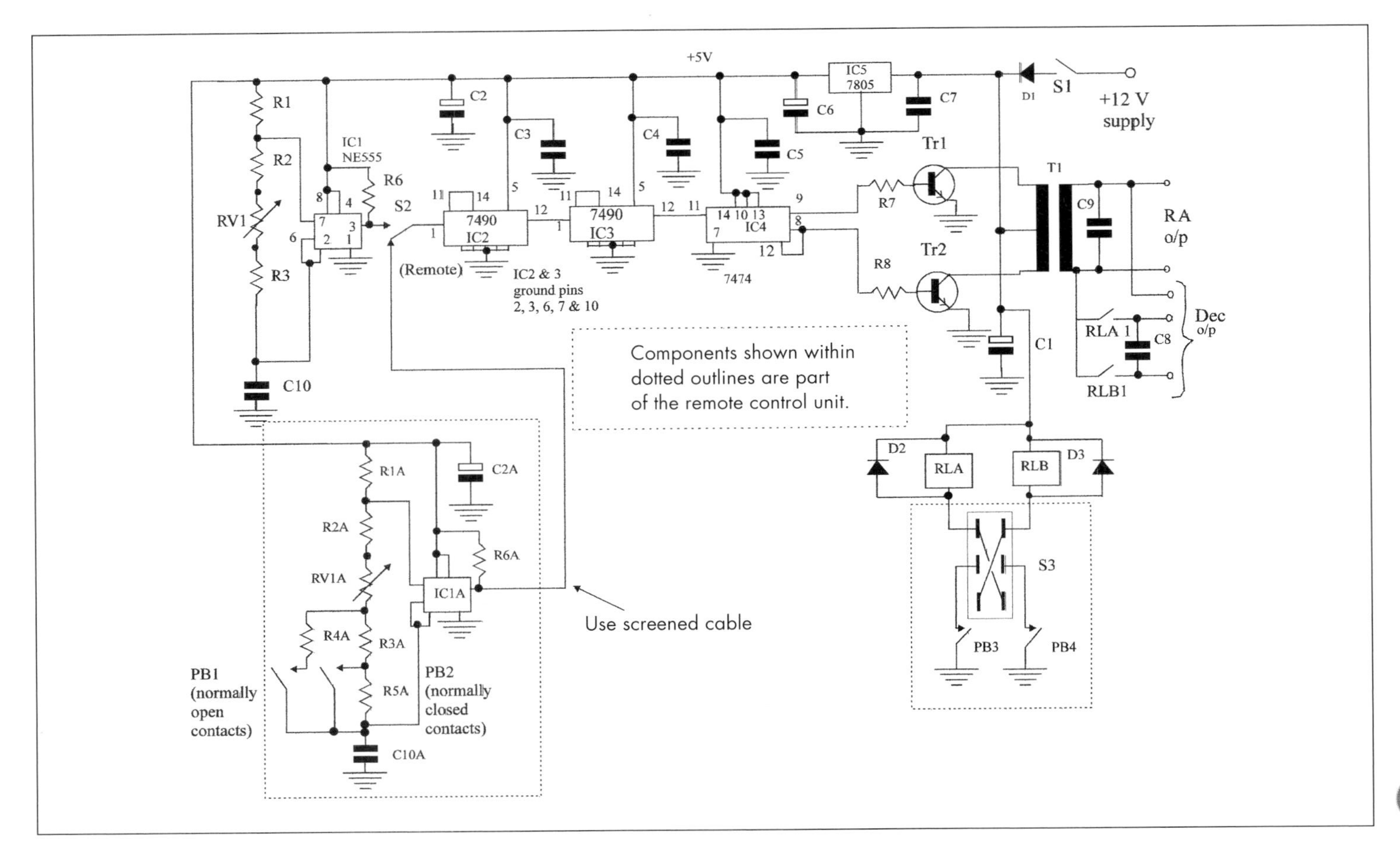
+5V
IC5 7805
S1
+12 V supply
D1
C2
C3
C4
C5
C6
C7
R1
R2
RV1
R3
C10
IC1 NE555
R6
S2
7490 IC2
7490 IC3
IC4
7474
(Remote)
IC2 & 3 ground pins 2, 3, 6, 7 & 10
Tr1
Tr2
R7
R8
T1
C9
RA o/p
Dec o/p
RLA 1
C8
RLB1
C1
Components shown within dotted outlines are part of the remote control unit.
D2
RLA
RLB
D3
S3
PB3
PB4
R1A
C2A
R2A
R6A
RV1A
IC1A
Use screened cable
R4A
R3A
R5A
PB1 (normally open contacts)
PB2 (normally closed contacts)
C10A

12 V supply to ground through transformer T1 that is a normal mains transformer connected in reverse.

The overall output frequency is adjustable (in the case of the 50 Hz version) from 47.5 Hz to 52.5 Hz by adjusting RV1, or from 40 Hz to 60 Hz by pressing PB1 or PB2. For use with 60 Hz motors, reduce the value of resistors R2 and R2A; their exact value is selected on test. A good starting point is 47 kΩ.

The switched declination output is taken from the output of T1 via the contacts of relays RLA or RLB, either of these being energised by pressing PB3 or PB4 on the remote unit. Capacitor C8 gives a phase shift to the declination motor. A value of 0.1 μF produces the correct phase change for a Crouzet motor, but may need changing for other types. Some confusion may arise over the use of a reversible a.c. motor. There are usually two sets of windings brought out to three connections. Supply is fed between the common connection and one winding, a series capacitor giving a phase-changed feed to the other. Wiring C8 across the output socket gives the desired effect. This capacitor *must be* 400 V working.

Note the declination reversing switch mounted on the remote control. This is used to reverse the "sense" of the pushbuttons to match the view through the eyepiece. A double-pole changeover switch wired from corner to corner will reverse the connection, as shown in Figure 15.1.

Use with 12 V d.c. declination motors requires a simple modification. The modification consists of the following:

> Omit the relay board completely. Wire a 12 V supply (taken from the junction of D1 and C1) and ground connection (from any ground point on the main board) through the S3, PB3 and PB4 combination to an output socket. Push-switches PB3 and PB4 need to be wired in a similar fashion to the "cross-connected" switch S3 – rather than their original simple connection. Ensure the declination motor's windings are isolated from ground, otherwise a short circuit is liable to occur. Pressing PB3 or 4 now supplies a positive and negative 12 V output that has reversed polarity depending on which button is pressed. Pressing both buttons together doesn't cause anything dramatic to occur; the motor only runs in one direction.

Construction

While it is possible to construct this project using Vero or similar perforated board, the use of a printed circuit board greatly aids reliability. Board layouts are available by Email from the author together with files suitable for printing via the shareware program "PIA". Ready-made boards are not available.

Start construction by fitting resistors and capacitors to the main board; solder the integrated circuits directly into the board. Avoiding the use of i.c. sockets will improve reliability in damp conditions. There are no special requirements for mounting other components. Take care to correctly insert the integrated circuits, transistors and electrolytic capacitors. Note that Tr1 and Tr2 are fitted with small heat sinks.

Figure 15.2 shows parts mounted on a sub-board. An offcut of copper-clad print board would be ideal. This makes mounting the pushbutton switches neater than fixing them to the front panel. Secure the remote oscillator board by soldering along one edge.

Use 5- or 6-core screened cable between the remote and main units. In the case of a 12 V declination motor, more cores will be required to accommodate the 12 V feed to the remote switches. Chromed brass, latching DIN connectors make a neat and tidy interface to the remote unit.

Testing

Caution:This unit produces an output equivalent to mains electricity – take care during testing.

Link the main board to give local operation (wire pin 3

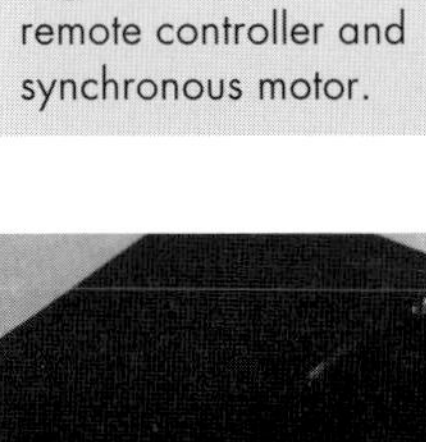

Figure 15.2 The remote controller and synchronous motor.

Figure 15.3 Controller and remote unit.

of IC1 to pin 1 of IC2). Mount RV1 on a small bracket and wire to the board. Connect the high-voltage (RA) output to a suitable motor. Use well-insulated connections and place the board securely on an insulated surface. Power the board from a fused 12 V source and check the current drawn, (around 500 mA). If the current differs widely from 500 mA, disconnect and check for missing components or shorts.

If a frequency counter is available, check the output frequency with a high-impedance probe on the low-voltage side of the output transformer. The frequency should be adjustable either side of 50 Hz (or 60 Hz). An alternative method would be to check the telescope drive against time. If the motor doesn't run, disconnect the motor and check the output voltage is between 200 and 230 V (100 to 110 V) a.c. The figures in brackets are for 110 V, 60 Hz motors. If the output voltage is correct without a load, the motor may be drawing more current than this inverter can supply.

The original Crouzet motors run quite well from this unit. Larger motors require more current; in this case it would be necessary to add an extra stage before the transformer.

Fitting It Into a Case

The exact size of case is not important. The original unit was constructed using a steel and aluminium case measuring 120 × 200 × 50 mm (5 × 8 × 2 in). Some constructors may wish to use a larger case and fit a battery inside. One additional item shown on the photographs is a mains neon wired across the RA output socket, which indicates high-voltage output and proves most useful. Various LED indicators can be added to show power; internal and external oscillators, and so on.

Once the unit has been thoroughly tested, it may be moisture-proofed by spraying both sides of the printed circuit boards with print circuit lacquer. Avoid spraying the switches.

Beware of adding too many bright indicator lights – they are distracting when taking photographs. Essential lights can be reduced in brightness by the careful use of black paint.

And Finally ...

In this modern age of low-voltage digital electronics, producing mains-level voltage to drive a low-power motor may seem rather quaint. Synchronous motors have an advantage over "modern" stepper motors of being virtually vibration-free. This prevents the mount resonating, producing star images as "blobs" or "streaks".

One word of caution: if you use a gel cell battery it should not be charged using a standard car battery charger. The correct type of charger can often be bought cheaply from radio-controlled model shops or from mail-order component suppliers.

Parts List for Telescope Controller

Parts suffixed 'A' are used in the remote oscillator board.
T1 230 V primary 9–0–9 V secondary 6 VA transformer (used in reverse)
N.B.: For 110 V motors, use a 110 V primary, 9–0–9 V secondary transformer.

IC1, IC1A NE555
IC2 and IC3 7490
IC4 7474
IC5 7805
Tr1 and Tr2 TIP41 (with clip-on heat sink)
D1, D2, D3 1N4001

C1 470 μF 25 V
C2, C2A 15 μF 16 V
C3, C4, C5,C7 0.1 μF 50 V min metallised polyester film
C6 47 μF 25 V
C8, C9 0.1 μF 400 V disc ceramic or polyester film
C10, C10A 820pF close-tolerance polystyrene

R1, R1A, R6, R7, R8 1 k R2, R2A 54 k (47 k for 60 Hz) select on test.
R3, R3A 27K
R4A,R5A 15K
RV1, RV1A 10K

All fixed resistors 1% metal film 0.3 W.

The 54 k resistor can be made from 51 k and 3 k3 resistors in series.

S1, S2 miniature single pole changeover (SPCO) toggle switch
S3 miniature double-pole changeover (DPCO) panel-mounting slide switch
PB1, PB2, PB3, PB4 DPCO pushbutton slide switch complete with mounting bracket
RLA, RLB 12 V coil, 240 volts a.c. rated contact miniature relay
Suitable mains-rated sockets and plugs for the output supply to the motors
Panel mount neon to indicate high-voltage output (mains-rated neon)
Panel mount fuseholder complete with 2 A fuse
Latching 5-way DIN plug and socket for connecting the remote unit (see text with reference to 12 V dec motors)

Gear wheel sets and motors can be obtained from Beacon Hill Telescopes, 112 Mill Road, Cleethorpes, DN35 8JD, UK. Tel +44 (0) 1472 692959.

Notes

Parts for the oscillator in the hand controller are identical to those used on the main board. Nothing in this design is critical; most-mail order component suppliers stock all the components and hardware. The frequency-determining components around IC1 need to be chosen with care to minimise drift due to changes in temperature. A good-quality potentiometer is required at RV1.

Part V

Appendices

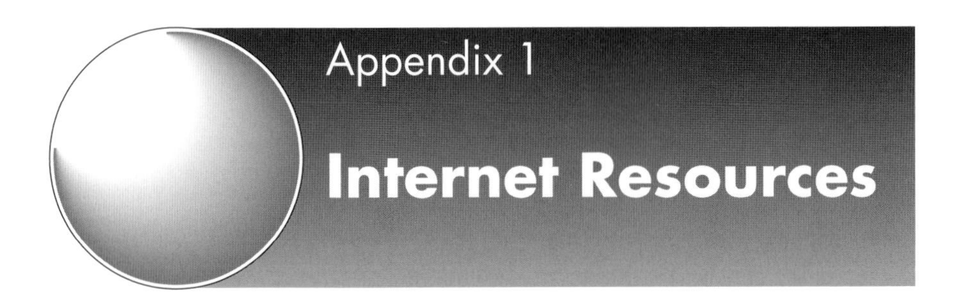

Appendix 1
Internet Resources

Dedicated Web Site

<http://www.springer.co.uk/astro/support/0007.html>

ATM Mailing List

To subscribe, send the line:
subscribe atm
in the body of an email to: <majordomo@shore.net>

The list is archived at:
<http://www.system.missouri.edu/ics/staff/andy/ATM/atmarchive.html>

Authors' Web Pages

Mel Bartels:	<http://www.efn.org/~mbartels>
David Johnson:	<http://www.astromag.co.uk>
Albert F. Kelly:	<http://www.ghg.net/akelly/>
Steven Lee:	<http://www.aao.gov.au/local/www/sl>
Euan G. Mason:	<http://www.fore.canterbury.ac.nz/EUAN.HTM >
Chuck Shaw:	<http://www.ghg.net/cshaw>
Stephen Tonkin:	<http://www.aegis1.demon.co.uk/atm.htm>
Scott Wilson:	<http://www.geocities.com/CapeCanaveral/Hall/3715>

General Resources

ATM Resources: <http://www.freenet.tlh.fl.us/~blombard/>

ATM FAQ (Frequently Asked Questions document):
<http://www.netacc.net/~poulsen/atm-faq.htm>
<http://www.aegis1.demon.co.uk/faq/atm-faq.htm>

Appendix 2
Contributors

Mel Bartels is a former professional musician who now works as an analyst/programmer. He develops software to aid ATMs. Mel runs the ATM Mailing List, where the ATM community exchanges innovative ideas and supports newcomers to the hobby.

Bratislav Curcic is an electronics engineer for a large telecommunication company. He started with astronomy *c.* 1974, almost immediately with telescope making as well. He moved to Australia from his native Yugoslavia in 1988.

David Johnson is a professional radio engineer currently employed in the public safety field. He has been an amateur astronomer for the past 30 years using various home-built bits and pieces, the largest of which is a 12 ft diameter observatory.

Albert F. Kelly is a technical contracts manager for NASA at the Johnson Space Centre (Houston, Texas) and has been an amateur astronomer since the mid-fifties. He built his first telescope in 1958 – a 4.25 in f/10 Newtonian made from carpet-roll tubing, duct tape, bits of stuff, and an optics set from Edmund Scientific. He has since been involved in the design and construction of about two dozen telescopes, ranging from 3 in to 32 in aperture. He built a CB245 *Cookbook* Camera in 1994, and has spent the last year or two trying to perfect personal techniques for tricolour CCD imaging.

Steven Lee is a night assistant at the Anglo-Australian Observatory where, in addition to "driving" the telescope, his duties include programming the telescope control system, designing (and sometimes making) optics for use on the AAT, and managing the archive of AAT data. He has been making telescopes since 1972 and now does his amateur observing from his home near Coonabarabran in the Australian outback.

Euan G. Mason is a senior lecturer in Forestry at the University of Canterbury, Christchurch, New Zealand, with professional interests in the mathematical modelling of tree growth, silviculture, and decision-support systems for forest managers. He enjoys all aspects of amateur astronomy, especially "creating gizmos with my own hands to make my observing more effective".

Terry Platt is an electronics engineer and Technical Director of a company which makes infra-red beam obstruction detectors. He is also the designer of the 'Starlight Xpress' range of CCD cameras for astronomy. Terry became interested in astronomy during the late 1950s, largely as a result of the beginning of the 'space age' and the appearance of Patrick Moore's series 'The Sky at Night'. He made a 6 inch Newtonian during 1961 and has built his own telescopes ever since. In recent years his interest in high resolution planetary work has led to experimentation with large off-axis reflectors. Terry was born in West Yorkshire, but has lived for many years in the Southern county of Berkshire, England. He is currently polishing the mirrors for a 14.5 inch 'Stevick-Paul' off-axis reflector and trying, yet again, to avoid getting pitch all over the living room carpet.

Chuck Shaw is Senior Flight Director for the Space Shuttle and Space Station Programs at NASA's Johnson Space Centre in Houston, Texas. His primary interests in amateur astronomy are telescope making and CCD imaging with the CB245 CCD camera he built. He is past President of the Johnson Space Centre Astronomical Society, where he actively works with both new and experienced observers. He also actively works with amateurs both in his own club and via the Internet to help them build their own scopes and tracking platforms.

Klaus-Peter Schröder is a research astronomer at the Institute of Astronomy at the Technical University at Berlin, from where he studies the atmosphere of red giants. He is an avid astrophotographer (contributor to *Handbuch der Astrofotographie*, Berlin, Springer 1996, edited by Bernd Koch), who takes his portable telescopes with him on his worldwide astrophotography expeditions.

Gary Seronik is an Assistant Editor at *Sky & Telescope* magazine in Cambridge, Massachusetts. A lifelong astronomy enthusiast and an avid observer, he credits his overwhelming desire to see intricate planetary detail with his passion for optimised optics and telescope making. While working at Vancouver's Pacific Space Centre, Gary conducted evening courses in mirror making and telescope basics.

Gil Stacy is a trial lawyer in Savannah, Georgia, who has been making telescopes since the age of fifteen. He divides his leisure time between astronomy and fly-fishing.

Stephen Tonkin is a peripatetic teacher of astronomy, in which he is able to combine his hobbies of astronomy, storytelling, drama and music. His interest in astronomy was triggered by a childhood spent under dark African skies and he delights in showing people how they can make serviceable kit from stuff that is normally considered to be junk.

Scott Wilson is a physics student at the University of South Carolina who has had an interest in binocular astronomy since his schooldays. He is past president of the Midlands Astronomy Club.

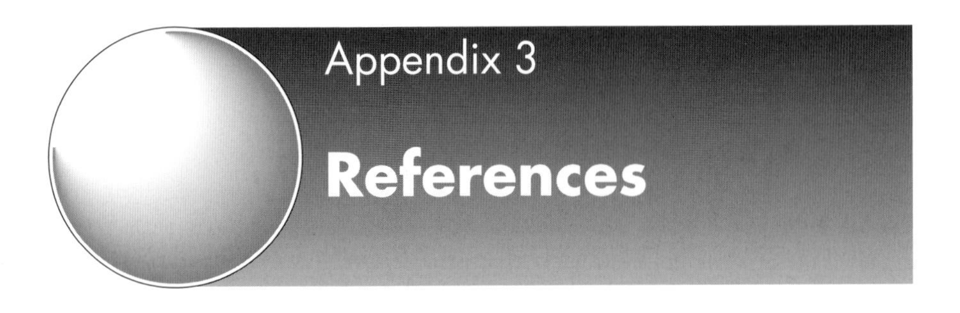

Chapter 6

Herrig E., "A new concept for tilted-component telescopes", *Sky & Telescope* (November 1997), 94(5): 113–115.

Chapter 7

De Vany A., *Master Optical Techniques*. New York: Wiley, 1981.

De Vany A., "A universal telescope of the Wright type", *Sky & Telescope* (August 1985), 70(2): 167–171.

Knott L., "Finish your mirror with polishing pads", *Sky & Telescope* (June 1992), 83(6): 688–690.

Rutten H and van Venrooij M., *Telescope Optics, Evaluation and Design*. Richmond, VA: Willmann-Bell, 1988.

Schmidt B., "A rapid coma-free mirror system", in Ingalls A (ed), *Amateur Telescope Making*; bk 3, New York, Scientific American, 1953.

Texereau J., *How to Make a Telescope*; 2nd edn. Richmond, VA: Wilmann-Bell, 1984.

Wright F B., "Theory and design of aplanatic reflectors employing correcting lens", in Ingalls A (ed), *Amateur Telescope Making*, bk 2, New York, Scientific American,1937.

Chapter 9

D'Autume G., "Equatorial table without a pivot", *Sky & Telescope* (September 1988), 76(3): 303–307

Poncet, A., "An equatorial table for astronomical equipment", *Sky & Telescope* (January 1977), 53(1): 64–67.

Chapter 10

Garlitz J., *Analysis Method for PEC Data:* <http://www.orednet.org/~jgarlitz/pec.htm>

Taki T., "A new concept in computer-aided telescopes", *Sky & Telescope* (February 1989), 77(2): 194–196.

Chapter 12

Grover R., *A Quartz Controlled Scotch Mount* <http://www.u-net.com/ph/mas/projects/scotch/scotch.htm>

Jones D., *Jones on Stepping Motors* <http://www.cs.uiowa.edu/~jones/step/>

Sinnott R., "Two arms are better than one", *Sky & Telescope* (April 1989), 77(4): 436–441

Trott D., "The double-arm barn-door drive", *Sky & Telescope* (February 1988), 75(2): 213–214.

Books

Berry R., *Build Your Own Telescope*, 2nd edn. Richmond, VA: Wilmann-Bell, 1994; ISBN: 0-943-39642-5.
Telescope projects suitable for the beginning ATM.

Berry R., Kanto V., Munger J., *The CCD Camera Cookbook: How to Build Your Own CCD Camera* (book and disk). Richmond, VA: Wilmann-Bell, 1994; ISBN: 0943396417.
All the information and software you need to make your own CCD camera.

Berry R., Kriege, D., *The Dobsonian Telescope: A Practical Manual for Building Large Aperture Telescopes*. Richmond, VA: Wilmann-Bell, 1997; ISBN: 0943396557.
Comprehensive guide to the subject from acknowledged experts in their field.

Brown S., *All About Telescopes*. Barrington, N.J., Edmund Scientific Co.; ISBN: 0-933346-20-4.
Very good beginners' reference.

Ingalls A.G. (ed.), *Amateur Telescope Making*, 3 vols. New York, Scientific American (out of print). Republished Richmond, VA: Willman-Bell, 1996; ISBN (Willmann-Bell editions): 0-943396-48-4 (vol. 1), 0-943396-49-2 (vol. 2), 0-943396-50-6 (vol. 3).
Excellent series, packed with information and advice.

Harrington P.S., *Star Ware*. New York, Wiley; ISBN 0-471576-71-9.
Not really an ATM book, but covers some useful projects.

Howard N.E., *Standard Handbook for Telescope Making*, 2nd ed. New York, Harper & Row, 1984; ISBN: 0-06-181394-X.
Perhaps the best book for the first-time telescope maker. Covers all aspects of making an 8 in f/7 plus a variety of mounts.

Mackintosh A. (ed.), *Advanced Telescope Making Techniques*, 2 vols. Richmond, VA: Willman-Bell, 1986; ISBN: 0-943396-11-5 (vol. 1) and 0-943396-12-3 (vol. 2).
Reprints from the Maksutov Circulars. *Worth having, but definitely not a beginner's book.*

Manly P.L., *Unusual Telescopes*. Cambridge University Press, 1991; ISBN 0-521-38200-9 (hardcover) or ISBN 0-521-48393-X (paperback).
Full of interesting and amusing ideas. Not a manual, but more a general interest book for armchair ATMing.

Miller R., Wilson K., *Making & Enjoying Telescopes – 6 Complete Projects & A Stargazer's Guide*. New York: Sterling, 1995; ISBN: 0-8069-1277-4.
Projects using bought components; no optical work is covered.

Rutten H., van Venrooij M., *Telescope Optics, Evaluation and Design*. Richmond, VA: Willmann-Bell, 1988; ISBN: 0-943396-18-2.
A "must have" for advanced workers.

Strong J., *Procedures in Applied Optics*. New York, Dekker, 1989; ISBN: 0824779878.
Design and construction of optical instruments.

Strong J., *Procedures in Experimental Physics* (also printed under the title *Physical Laboratory Practice*). Bradley, IL, Lindsey, 1938; ISBN: 0-917914562.
Part of the book is on laboratory optical work.

Suiter, H.R., *Star Testing Astronomical Telescopes, A Manual for Optical Evaluation and Adjustment*. Richmond, VA: Willmann-Bell, 1994, ISBN: 0-943396-44-1.
This is THE book on star-testing.

Texereau J., *How to Make a Telescope*, 2nd edn. Richmond, VA: Willmann-Bell, 1984; ISBN: 0-943396-04-2.
Considered by many to be the best book for general ATM optical work.

Thompson A.J., *Making Your Own Telescope*. Cambridge, Mass., Sky Publishing, 1947, revised, 1973; ISBN: 0-933346-12-3.
Detailed instructions on how to make a 6 in f/8 equatorial Newtonian.

Trueblood M., Genet R., *Telescope Control*. Richmond, VA:Willmann-Bell, 1997;ISBN: 0943396530.
Control of telescopes using microcomputers.

Periodicals

ATM Journal
17606 28th Ave. S.E.
Bothell, WA 98012, USA
Tel: (206) 481-7627
Web: <http://www.halcyon.com/rupe/atmj/>
Since the demise of Telescope Making, *this has been the only dedicated ATM periodical.*

Amateur Astronomy
3544 Oak Grove Drive
Sarasota, Florida 34243, USA
Tel/fax: (941)355-2423

Sky & Telescope
PO Box 9111
Belmont, MA 02178-9111, USA
Tel: (800) 253-0245 (US and Canada orders by credit card)
all others call +1-617-864-7360
Web: < http://www.skypub.com>

Astronomy Magazine
21027 Crossroads Circle
PO Box 1612
Waukesha, WI 53187, USA
Tel: (800) 533-6644
Web: <http://www.kalmbach.com/astro/astronomy.html>

Observatory Techniques Magazine
1710 SE 16 Ave
Aberdeen, SD 57401-7836, USA
Tel/Fax: 605-226-1078